Maintenance Strategy

To the management of Australian industry for providing much
of the information that allowed this book to be written and
for their hospitality on my many visits to their country.

Maintenance Strategy

Dr Anthony Kelly
Principal Consultant, IMMS
Visiting Professor at Central Queensland University, Australia
Stellenbosch University, South Africa
Høgskolen i Stavanger, Norway

Butterworth-Heinemann
Linacre House, Jordan Hill, Oxford OX2 8DP
A division of Reed Educational and Professional Publishing Ltd

 A member of the Reed Elsevier plc group

OXFORD   BOSTON   JOHANNESBURG
MELBOURNE   NEW DELHI   SINGAPORE

First published 1997
© Reed Educational and Professional Publishing 1997

British Library Cataloguing in Publication Data
A catalogue record for this book is available from the British Library

ISBN 0 7506 2417 5

Library of Congress Cataloguing in Publication Data
A catalogue record for this book is available from the Library of Congress

Printed and bound in Great Britain by
Biddles Ltd, Guildford and King's Lynn

Contents

Preface

Devising optimal strategy for maintaining industrial plant can be a difficult task of quite daunting complexity. My aim, therefore, in writing this book, has been to provide the plant engineer with a comprehensive and systematic approach for tackling this problem, i.e. a methodology — or framework of guidelines—for deciding maintenance objectives, formulating equipment life plans and plant maintenance schedules, designing the maintenance organization and setting up appropriate systems of documentation and control. I have called this approach Business-centred Maintenance (BCM) because it springs from, and is driven by, the identification of business objectives, which are then translated into maintenance objectives and underpin the maintenance strategy formulation.

I have developed this approach during more than twenty years' full-time involvement with maintenance management — teaching it (mostly in-plant), researching its complexities and (more recently) providing industrial consultancy in the subject. Indeed, it is the last of these activities that has had the greatest influence on the content of this book, leading me to modify and expand the approach outlined in my previous books* and enabling me to illustrate it with industrial examples and case studies derived from my own work.

Currently, the BCM methodology is being used, by my own partnership** and by other consultancy groups, to audit the maintenance departments of industrial companies with a view to their modification and improvement. It has been adopted as the framework for postgraduate programmes in maintenance management, at the Universities of Manchester (UK) and of Central Queensland (Australia). Parts of it, e.g. the top-down-bottom-up formulation of plant life plan and preventive schedule (see Chapter 9) have been incorporated in maintenance management software (available from MMS, Adelaide and from Mechatricity, Brisbane, Australia).

Most publications in this subject area have lacked the analysis of maintenance management principles and structures that is essential for the development of the topic as an academic discipline. I hope that in trying to correct this situation I have provided a book that will help not only students of industrial management but also practising maintenance managers.

---

* *Management of Industrial Maintenance* (with M. J. Harris), Newnes-Butterworths (1978)
   Maintenance Planning and Control, Butterworths (1984)
   Maintenance and its Management, Conference Communication (1989)
** International Maintenance Management Specialists (IMMS): A. Kelly, M. J. Harris, H. S. Riddell, A. D. Ball (Associate), P. Bulger (Associate), and T. Lenahan (Associate).

Chapter 1 takes the systems view of a company and explains that the maintenance sub-system influences — and is influenced by — many other sub-systems. It emphasizes that the capital asset management function has a major effect on the maintenance department via its concern for asset reliability and maintainability and also that, as regards organizational design, the maintenance and production departments are inseparable. Chapter 2 looks at the influence of capital asset acquisition policy on maintenance life-cycle costs. Via an industrial case study, Chapter 3 then develops the overall methodology of BCM. Chapter 4 shows how an industrial plant can be modelled as a hierarchy of inter-related parts and also as a process flow. Chapters 5 and 6 then explain how statistical techniques can be used first to model patterns of component failure and quantify component reliability and secondly to model and assess the reliability of plant systems. As well as showing how business needs determine the development of maintenance objectives Chapter 7 also outlines a hierarchy of such objectives. Chapter 8 then deals with what is probably the key issue in this area, namely preventive maintenance decision-making, discussing the concepts and principles involved and their application to the formulation of a life plan for a unit of plant. Chapter 9 outlines the unique TDBU approach to formulating a preventive maintenance schedule for a plant and Chapter 10 then describes a reliability-based model for controlling the application of maintenance effort. To further illuminate the ideas which have been discussed up to that point, and to reinforce understanding of them, Chapters 11 and 12 present various contrasting industrial case studies and exercises. Finally, Chapters 13, 14 and 15 first review the merits and limitations of the two other basic philosophies of maintenance strategy formulation, namely Reliability centred Maintenance (RCM) and Total Productive Maintenance (TPM) and then compare and contrast them with BCM.

This book deals with the general BCM methodology for deciding maintenance strategy, i.e. with the setting of objectives and with the determination of life plans for the various units of an industrial plant and of schedules for the complete plant. A companion book, *Maintenance Organization and Systems*, deals with the concomitant organizational and control elements of maintenance management.

Acknowledgements

I am deeply indebted to my colleague John Harris who most generously contributed Chapters 5, 6 and 13 and also edited the complete text. I would also like to thank Dr Andrew Ball who contributed Table 8.3 and Dr Harry Riddell for the main figures in Chapter 2.

The following have also contributed through discussion and correspondence arising out of my own and my IMMS partners' industrial consulting work:

John Abbott and Brian Gover, Comalco, Australia
Jim Beckford, Mars Confectionery, UK
Alan Bonney, BHP Coal, Australia
Colin Bower, ex-QEC, Australia
Tony Calloway, Cummins, UK
Bill Carr, Alcan, UK
Glen Chuter, Alcan, Australia
Harry Cockerill, Foster-Wheeler, UK
Alan Dundass, Nabalco, Australia
David Eiszele, John Collins and Bill Wallace, Western Power, Australia
Richard Grey, Courtaulds Chemicals, UK
Bent Knauer, Bignum and Stenfor, Denmark
Nigel Land, Conoco, UK
Peter Mackenzie, MINCOM, Australia
Jeff Miller, Peak Gold Mines, Australia
David McLatchie, Petroleum Refineries (Australia)
Tom Muldoon and Dermot Connellan, ESB, Ireland
Jerry Murden, Du Pont Algraphy, UK
Peter Mackenzie, MINCOM, Australia
Ray Parkin, Capcoal, Australia
Norman Peacock, ICL, UK
Ian Roberts, ECNZ, New Zealand
Liam Tobin, Boyne Smelters, Australia
Barry Wilmer, Nissan, UK
Mark Zamitt, QAL, Australia

I would also like to thank:

- Bill Geraerds, Emeritus Professor of Industrial Engineering, University of Eindhoven, Holland, for help and advice which has greatly influenced my work;
- Christer Idhammer, Idcon, USA, for his insights into fundamental maintenance concepts;

- Professor Floyd Miller, University of Illinois, USA, for his assistance with my research on fleet maintenance;
- John Day, Alumax, USA, for informative discussions regarding his own approach to maintenance management;
- the many former students at the University of Manchester's School of Engineering whose research projects provided much of the information on which this book is based — in particular John Halstead, Greville Seddon, Chris Bull, Mohammed Al-Fouzan, Julia Gauntly and the various postgraduates seconded from the ESB, Republic of Ireland

It has been my experience in presenting many industrial training and educational courses in maintenance management which prompted me to put this book together and I am therefore also indebted to those who have helped in the organization of these, notably David Willson of Conference Communication, Michael Shiel of the Irish Management Institute, Len Bradshaw of EIT (Australia) and, in particular, Professor Erin Jancauskas, Dean of Engineering at Central Queensland University, Australia (for having the foresight and enthusiasm to set up the CQU postgraduate distance-learning degree in maintenance management).

Finally I have to thank Denise Jackson, Vicky Taylor, Carol Critchley and Beverley Knight, of the University of Manchester School of Engineering, for producing the figures and typescript, and especially for their remarkable forbearance when dealing with my countless alterations.

Anthony Kelly
IMMS
Bollin House
5 Edgeway
Wilmslow SK9 1NH
Cheshire UK
Fax 01625 539585
Tel 01625 529379

1
Maintenance and the industrial organization

Organizations and the role of management

Etzioni defined organizations as groupings of human beings (of individuals and sub-groups of individuals) constructed and reconstructed to seek specific goals[1]. Various material resources will also be needed, he said. A better understanding of organizations may be obtained through the so-called *systems* approach. In this, organizations can be viewed as open systems taking inputs from their environments and transforming them — by a series of activities and with some objective in view — into outputs (see Figure 1.1).

Organizations can be categorized, on the basis of their objectives, into public and private enterprises. An industrial company exemplifies the latter and Riddell has pointed out that if it is to achieve its primary objective of maximizing its long-term profitability — while also providing an in-demand service — it will need to carry out two prime functions[2]:

- First, the internal mechanisms of the industrial enterprise itself must be made to operate well. The right product must be made at the right time, by the right plant, using the right raw materials and employing the appropriate workforce. The physical assets must be carefully selected and properly maintained. Effective long-term research and development plans must be implemented and new capital investment generated. In short, the internal efficiency must be high.
- Secondly, the interaction with the outside world, with external influences and constraints, must be made to be co-operative and beneficial, rather than antagonistic and damaging, i.e. the overall, externally measured, efficiency must also be high.

Riddell sees the 'role of management' as being concerned with carrying out these functions in order to ensure the ongoing success (profit) of the organization. He sees management as the designer, constructor, director and controller of the organization so that it can achieve its objective. Several helpful approaches to carrying out this role have evolved (see Table 1.1). These, in particular the 'administrative' and the 'human relations' approaches, will be used in this book to develop a framework (or methodology) of maintenance management principles and procedures (see Chapter 3).

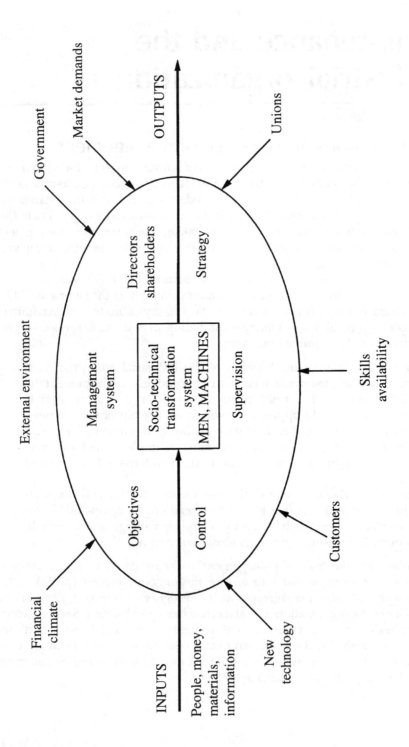

Figure 1.1. An organization as an open system

Table 1.1. Summary of management theories

Mechanistic management — Monitors and controls the way the job is performed at shop floor level; includes method, timing and direction.

Administrative management — Applies universal management functions and structural principles to the design of an organization and to its operation.

Human relations management — Studies characteristics and relationships of individuals and groups within an organization and takes account of these factors when designing and administering the organization.

Decision management — Applies procedural and quantitative models to the solution of management problems. A theory for communications and decision-making in organizations.

Systems management — Studies organizations as dynamic systems reacting with their environment. Analyses a system into its sub-systems and takes account of behaviourial, mechanistic, technological and managerial aspects.

Contingency management — Takes the view that the characteristics of an organization must be matched to its internal and external environment. Since these environments can change it is important to view the organizational structure as dynamic.

A systems view of maintenance management

Several writers have modelled the industrial organization as a socio-technical system comprising various sub-systems. For example, Kast and Rosenzweig saw it as an open, socio-technical system (see Figure 1.2) with the following five sub-systems, each with its own input–conversion–output process related to, and interacting with, the other sub-systems[3]:

(i)   a goal-oriented arrangement; people with a purpose,

(ii) a technical sub-system; people using knowledge, techniques, equipment and facilities,

(iii) a structural sub-system; people working together on integrated activities,

(iv) a psycho-social sub-system; people in social relationships, co-ordinated by

(v)   a managerial sub-system; planning and controlling the overall endeavour, i.e. ensuring that the activities of the organization as a whole are directed towards the accomplishment of its objectives.

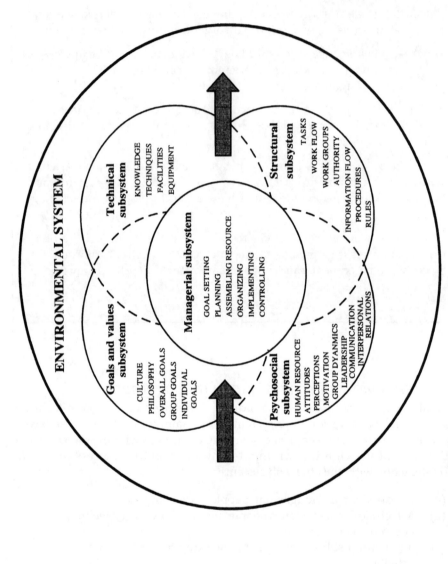

Figure 1.2. The organizational system

The author prefers to view the industrial organization as an open system, converting raw material or information into finished products of a higher value. It can be considered to be made up of many interacting sub-systems, e.g. maintenance, production, stores, capital asset acquisition, safety, design, finance, corporate finance, each carrying out distinct organizational functions.

- The function of *corporate management* (the master sub-system) is to set the organizational goal and strategy and direct, co-ordinate and control the other sub-systems to achieve the set goal.

- The function of *capital asset acquisition* is to select, buy, install and commission physical assets, a function which is carried out through the combined efforts of a number of other sub-systems, e.g. design, finance, projects.

- The function of *maintenance* is to sustain the integrity of physical assets by repairing, modifying or replacing them as necessary.

Each such sub-system requires inputs of information and resources from one or more of the other sub-systems and/or the external environment in order to perform its function. The output from one sub-system can be an input to another or an output to the external environment (see Figure 1.3).

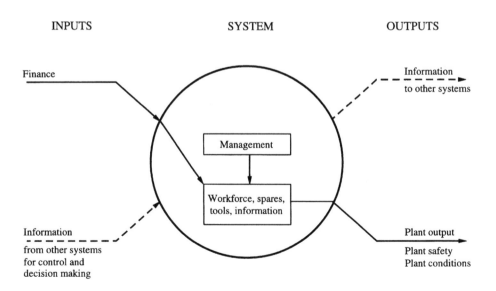

Figure 1.3. The function of the maintenance system

This systems view of an organization shows that the maintenance sub-system influences — and is influenced by — many of the other sub-systems. Two aspects of this are of particular importance:

(i)   Because the asset acquisition function — which influences reliability and maintainability — has a considerable effect on the maintenance function a necessary preliminary to developing the main arguments of this book is to clarify the nature of the relationship between them (see Chapter 2).

(ii) The relationships between maintenance and the other organizational sub-systems, e.g. production, must also be clarified, and must form part of any description of the operation of the maintenance sub-system or of any of its parts (see Chapter 3).

References

1.   Etzioni, A., *Modern Organisations*, Prentice Hall, Englewoood Cliffs, NJ, USA, 1964.
2.   Riddell, H. S., *Lecture Notes on Engineering Management*, University of Manchester School of Engineering, 1994.
3.   Kast, F. E. and Rosenzweig, J. E., *Organisations and Management*, 3rd edn, McGraw-Hill, Singapore, 1974.

2

Plant acquisition policy and maintenance life-cycle costs*

Life-cycle costing

One way of considering the profitability of plant is on the basis of its complete *life-cycle*. Figure 2.1 models the principal phases of this, and Table 2.1 lists the main cost-influencing factors. The importance of these various phases and factors will vary with the technology concerned, e.g. in power generation fuel costs may be the overriding factor, in petroleum refining the plant availability, in the provision of buildings their anticipated useful life.

Investment in the plant occurs from its conception to its commissioning, and perhaps into its early years of operation. If all goes well, the return on this investment begins soon after the plant comes into use and continues until the plant is disposed of. An example of a life-cycle cost profile is shown in Figure 2.2.

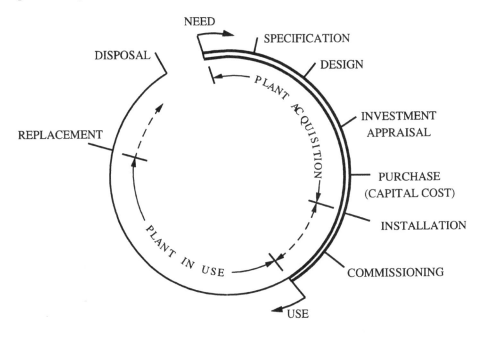

Figure 2.1. Plant life-cycle and costs

* Table 2.2 and Figures 2.2, 2.3, 2.4, 2.6 and 2.7 have been contributed by Dr H. S. Riddell, formerly of the University of Manchester School of Engineering.

Table 2.1. Factors influencing life-cycle profitability

ACQUISITION COSTS
Capital cost
Installation cost and time
Commissioning cost and time

OUTPUT PARAMETERS
Useful life
Plant performance
Product quality
Plant availability

RUNNING COSTS
Production cost
Maintenance cost
Fuel cost

OUTSIDE MANAGEMENT CONTROL
Product demand
Product price
Obsolescence

The data of this example have been used to plot Figure 2.3, which demonstrates that in some cases the total maintenance cost can be considerably greater than the capital cost.

A company might have as its objective the maximization of its plant's life-cycle profitability within the constraints imposed by the need for safe operation. Achievement of this would necessitate, among other things, an investment appraisal which sought an economic compromise between such factors as capital cost, running cost, performance, availability and useful life. Figures 2.4(a) and (b) illustrate the use of a life-cycle cost analysis to assist in such an investment appraisal decision.

Capital asset management

Almost invariably, the application of life-cycle cost analysis is rendered difficult by:

  (i)   the lack of definition of the capital asset acquisition sub-
        system;
  (ii) the complex relationships between the many factors
        involved in the economic compromise;
  (iii) the uncertainty of much of the life-cycle information, i.e.
        concerning such matters as the projected need for the

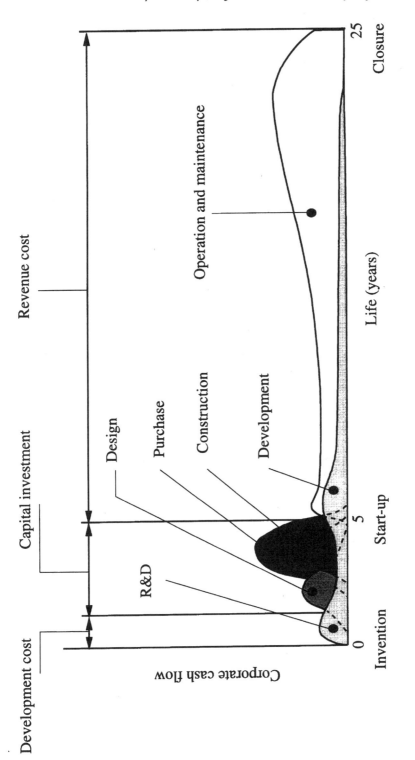

Figure 2.2. A life-cycle cost profile © *H. S. Riddell*

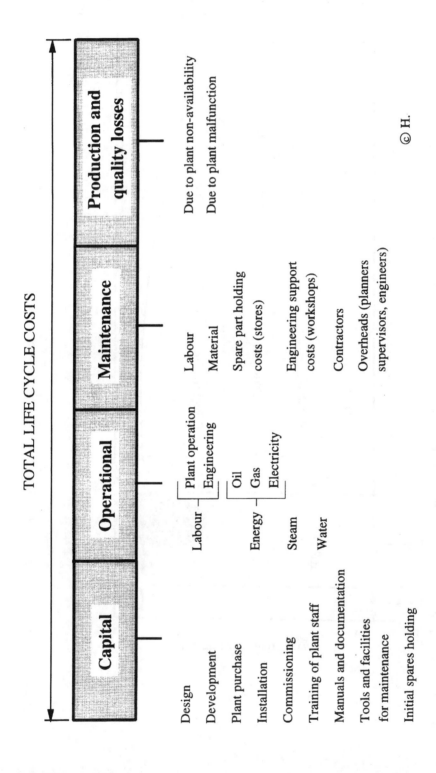

Figure 2.3. Factors in the build-up of total life costs   © H.S. Riddell

product, whether and when the plant would become obsolescent, the estimated plant reliability and availability, anticipated costs, and so on.

Because of these and other difficulties the equipment acquisition appraisal is usually dominated by considerations of plant performance and capital cost. Little or no thought is given to reliability and maintainability, the inevitable consequence being that installation and commissioning times and costs will be extended and that plant operation will be dogged by low equipment availability (i.e. high maintenance costs, both indirect and direct).

The question therefore arises as to how this situation can be improved or corrected. Is it via the so-called *terotechnological* approach? This evolved in the UK in the early 1970s. It was defined, at first, as follows[1]:

- A combination of management, financial, engineering and other practices applied to physical assets in pursuit of economic life-cycle costs

A little later, the following was added:

- .. its practice is concerned with the specification and design for reliability and maintainability of plant, machinery, equipment, buildings and structures, with their installation and replacement, and with the feedback of information on design, performance and costs.

In short, the idea quite rapidly enlarged from being an approach in which maintenance and unavailability costs were of central importance to one which was much more general, and therefore less tangible. Because of this the concept never took root in British industry.

Capital Asset Management, outlined in Table 2.2, is a more recent approach — preferred by the author — to this area[2]. It is based on the idea of 'optimizing total maintenance costs over the equipment life-cycle'. This is best achieved through an understanding of the effects that decisions taken in the plant's *pre-operational* phases can have on the direct and indirect maintenance costs of the *operational* phase[3] (see Figure 2.5).

The *specification* for new plant must include requirements for reliability and maintainability (i.e. availability) as well as for performance, capital cost and safety. As far as possible, the expected or useful life of the plant should also be specified. In support of this the equipment manuals, drawings, spares lists, spares security-of-supply and training needs should all be specified and, where necessary, this should be included in the contract.

At the *design* stage, reliability, maintainability and useful life are of para-mount importance and should be considered alongside performance. The method of production is particularly important. For example, if a continuous rather than a batch process is adopted, careful consideration should be given to the much higher maintenance costs that inevitably occur. In addition, it

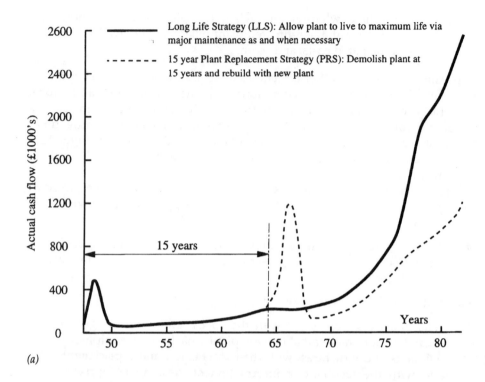

(a)

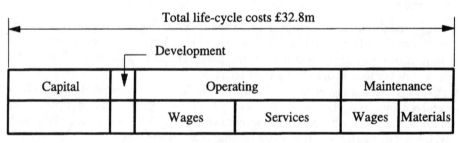

Present value costs, LLS (existing strategy)

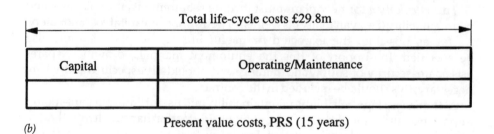

(b)   Present value costs, PRS (15 years)

Figure 2.4. Comparison of (a) actual costs and (b) present value costs for LLS and PRS, batch chemical plant

Table 2.2. Capital Asset Management

Definition	Practice	Implementation
A co-ordinated management of the design, procurement, use and maintenance of a firm's fixed assets, in order to maximize the contributions to the firm's profit over the life-cycle of those assets	Is concerned with: • the specification and design for reliability and maintainability of plant, equipment, buildings and services • their installation, commissioning, maintenance, modification and replacement • feedback of information on design, performance and costs	**1.** Correctly specify, design and acquire the asset **2.** Use the resources efficiently **3.** Determine and provide the appropriate level of care through effective maintenance **4.** Determine the optimum replacement periods

must be understood that design-stage considerations of reliability and maintainability can also affect the duration and cost of commissioning. It is self-evident that quality control during the *plant manufacture* stage will strongly affect the subsequent level of maintenance.

At the *installation* stage, maintainability will continue to be an important consideration because it is only then that the multi-dimensional nature of many of the maintenance problems becomes clear.

The *commissioning* stage will not only be a period of technical performance testing but also one of learning — where primary design faults, that might reduce availability, might be located and how they could be designed out. Failure to do this will mean serious maintenance problems and high unavailability early in the operational life. Operating equipment past its useful life stage will result in low availability and high maintenance costs.

Clearly, the best time to influence maintenance and unavailability costs is before the plant comes into use (see Figure 2.6).

- The opportunity for maintenance cost reduction is high at the design stage but drops rapidly — via several key, gateway decisions — to a relatively low level after commissioning.
- It is important that the often conflicting requirements of non-maintenance departments (represented, in Figure 2.6, by the downward-pointing arrows) are balanced against the maintenance requirements (represented by the upward-pointing arrows).

The above arguments suggest the following rules for the effective application of the Capital Asset Management, life-cycle approach to maintenance management.

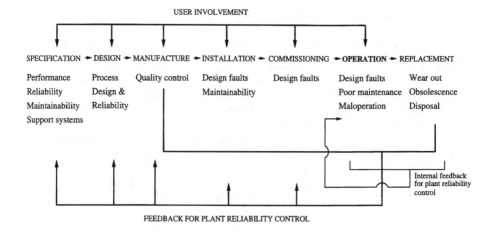

Figure 2.5. Factors influencing maintenance costs over the life-cycle

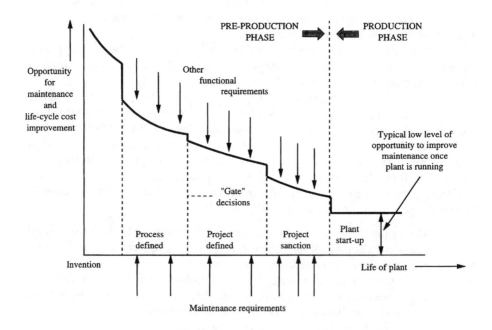

Figure 2.6. Opportunities for maintenance and life-cycle cost improvement © H. S. Riddell

(i)   Decisions to buy new or replacement plant should be based on a present-value life-cycle analysis of costs which should consider both maintenance and unavailability costs, these being estimated, wherever possible, from documented experience.

(ii) The owner–operator of the plant should co-operate with the designer–manufacturer–installer in a full analysis of its reliability, maintainability and safety characteristics. Such an exercise should include assessment of spare part provisioning, of maintenance personnel training and of supplier support systems. The higher the potential costs of maintenance and unavailability the more vital is this exercise.

(iii) The owner–operator should set up a system to record and analyse plant failures and identify areas of high maintenance cost. Such a system should operate from commissioning (with the supplier's assistance) to plant replacement. It should identify causes and prescribe solutions with the aim of minimizing the total of direct and indirect maintenance costs. Because plant design is a continuing process, information thus gathered should, ideally, be fed back to the equipment supplier or manufacturer. In certain circumstances it could be fed further to a data bank shared on an inter-company, national or inter-national basis. (The difficulty of implementing such information feedback continues to pose a major obstacle to the successful implementation of Capital Asset Management; communication systems are expensive and different organizations, with their different objectives, are involved during the equipment life-cycle.)

(iv) Within the organization concerned a Capital Asset Management system should be defined and established. This should transcend traditional functional boundaries for decision-making and will require considerable commitment from the most senior management for its effective operation. A model of such a system is shown in Figure 2.7.

Summary

The application of the Terotechnological/Capital Asset approach involves much higher expenditure than the traditional lowest-bid, lowest-cost, shortest-time approach. The difficulties of its implementation are many — e.g. cash constraints, time constraints, the uncertainty of forecasting demand and product life — so in some situations it has to be accepted that the extra effort and cost might not be worth the return. However, with the present trend towards automated, large, expensive plant, the adoption of this approach will usually bring considerable benefits. It requires the commitment and foresight of the most senior management. It is therefore no accident that the successful industrial examples of its application appear to have one common factor, at least — an engineering

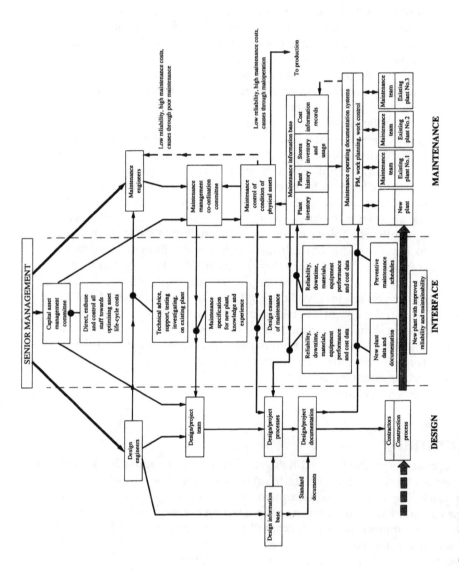

Figure 2.7. Capital asset management control system   © *H. S. Riddell*

director who is convinced of the long-term advantages of keeping maintenance firmly in mind when designing, installing and commissioning.

The maintenance manager's basic task is twofold, the determination of strategy and the organization of resources (i.e. labour, material and tools). These are difficult and important tasks in their own right, but failure of the organization to appreciate the fundamental ideas of Capital Asset Management will probably mean that the maintenance manager will be wasting his time on unnecessary tasks when the plant comes into operation.

References

1. Committee for Terotechnology, Department of Industry, Terotechnology, an Introduction to the Management of Physical Resources, HMSO, 1975.
2. Riddell, H. S., Life cycle costing in the chemical industry, *Terotechnica* **2**(1), pp. 89–104, 1980.
3. Kelly, A., *Maintenance Planning and Control*, Butterworths, 1984.

3
Formulating maintenance strategy, a business centred approach

The maintenance system

A methodology for developing a maintenance strategy is outlined in Figure 3.1. It is based on well-established administrative management principles (see Figure 3.2), provides a framework for identifying, mapping and then *auditing* the elements of any maintenance management system, and is called by the author the *Business Centred Maintenance* (BCM) approach.

One way of describing the *function* of a maintenance department is to say that it is 'ensuring and controlling the reliability of the plant'. The ways in which this function might be affected by its dynamic relationship with the production

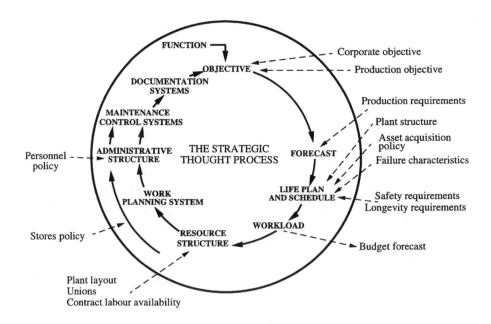

Figure 3.1. A methodology for developing maintenance strategy

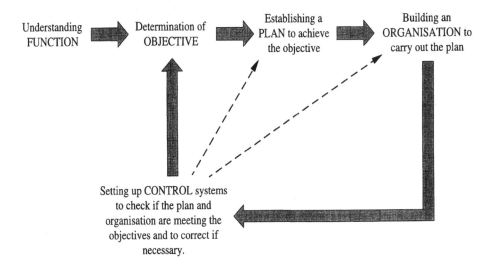

Figure 3.2. The basic steps of the management process

system need to be clearly understood. Once this has been achieved a definition of the maintenance *objective* that is compatible with the corporate and the production objectives can be identified. This might well be, for example, as follows:

> to achieve the agreed plant operating pattern and product quality, within the accepted plant condition and safety standards, and at minimum resource cost.

Consider, for example, the Food Processing Plant (FPP) outlined in Figures 3.3 and 3.4, which show the layout and the process flow. It operates for fifteen continuous shifts per week, Monday to Friday. It is the responsibility of the FPP users to specify the product mix and output (in cans/week) they desire, and hence the maximum allowable downtime — and also various quality, safety and plant longevity requirements. The maintenance department is responsible for ensuring that — at minimum resource (labour, materials, tools) cost — the plant is capable of meeting these requirements. The maintenance objectives need to be interpreted in a form that is meaningful at the main equipment level (that of a mixer, say, in the case of the FPP). This allows the maintenance *life plan* for the various units of plant to be established — on the basis of which the maintenance *schedule* for the plant is negotiated with production, taking into consideration the way in which the plant is used (the production policy often drives the maintenance schedule) and the level of plant redundancy. The main

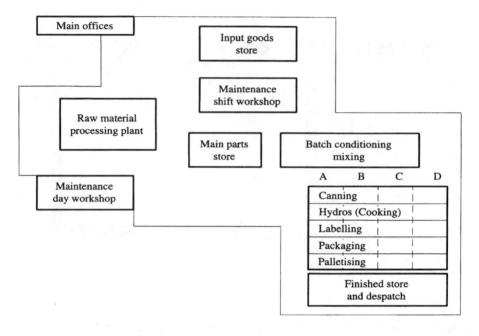

Figure 3.3. Food processing plant layout

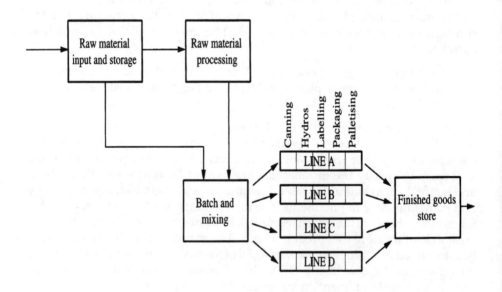

Pattern of operation 50 weeks x 5 days x 3 shifts, Monday/Friday

Figure 3.4. Food processing plant flow

decision regarding the life plan is the determination of the nature and intensity of the preventive work — which, in its turn, determines the resulting level of corrective work. The philosophy on which the FPP's maintenance life plan and schedule is based is outlined in Table 3.1.[1]

Table 3.1. Food processing plant, maintenance philosophy

Time period	Maintenance philosophy	Work type
Monday to Friday	'Keep the plant going' and	Reactive maintenance
	'Keep an eye on its condition'	Operator monitoring Tradeforce line-patrolling Condition-based routines
Weekends	'Inspect the plant carefully and repair as necessary in order to keep it going until next weekend'	Known corrective jobs Inspect and repair jobs Fixed-time jobs
Summer shutdown	'Carry out major jobs to see us through another year'	Known corrective jobs Fixed-time major jobs

The maintenance life plan and schedule strongly influence the level and nature of the *workload* (of preventive, corrective and modification work). The latter can be mapped by its organizational characteristics, i.e. its scheduling lead time, and for the FPP this is as shown in Figure 3.5. It can be seen that the workload — when considered alongside such factors as plant layout — has a considerable influence on the nature and design of the maintenance *organization*.

The primary task of the maintenance organization is to match resources to workload, and in so doing to ensure that the agreed plant output is achieved at minimum resource cost — which is a re-statement of the maintenance objective. In order to achieve this the organizational design needs to be aimed at maximizing tradeforce performance — which itself is a function of trade-force utilization and motivation, of the availability of spares, tools and information and of work planning efficiency. Many inter-related decisions have to be made (Where to locate the manpower? How to extend inter-trade flexibility? Who should be responsible for spare parts? Who should be responsible for maintenance information?) each influenced by various conflicting factors. Thinking in terms of the framework of Figure 3.1 reduces the complexity of this problem, by categorizing the decisions according to the main elements of the organization, namely its resource structure, administrative structure, work planning system and so on, and then considering each element in the order indicated. The procedure is iterative.

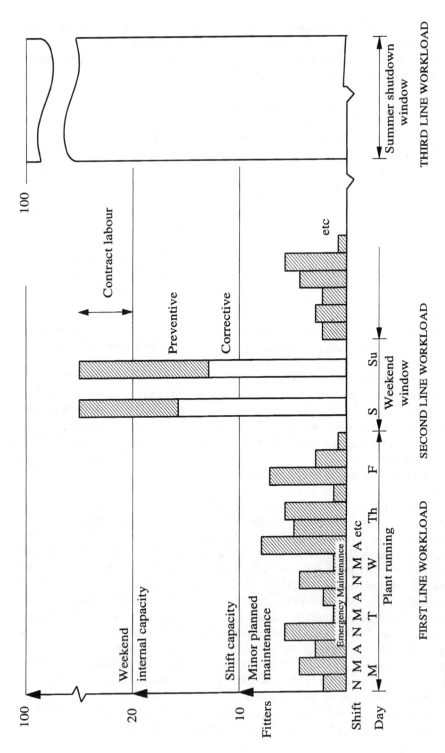

Figure 3.5 Workload pattern for fitters

The *resource structure* is the geographical location of workforce, spares, tools and information, their function, composition, size and logistics. For the FPP, for example, Figure 3.6 shows the Monday-to-Friday structure that has evolved to best suit the characteristics of a 24-hour first-line reactive workload. The emphasis is on rapid response, plant knowledge via specialization, shift working, and team working with Production. The shift groups have been sized to match the reactive workload. Lower priority jobs are used to smooth the workload.

Figure 3.7 shows the structure that matches the second-line weekend workload. This work is made up of relatively small preventive and corrective jobs that benefit from planning and scheduling. Contract labour is used to top-up, as necessary, the internal labour force. A similar approach is used for the annual shutdown, but in that case the contracted workforce exceeds that of internal labour.

The example shows the influence of the workload on the form of the organization. In practice, the purpose of any resource-structure design or modification will need to be specified and will most likely be expressed as being 'to achieve the best resource utilization for a desired speed of response and quality of work'. In general, this comes down to determining the appropriate shape (first line/second line/third line trade groups) and size of the trade groups.

The second element in the design of a maintenance organization is the formation of a decision-making structure — the maintenance *administrative structure*. This can be considered as a hierarchy of work roles, ranked by authority and responsibility, for deciding what, when and how maintenance work should be carried out. An example is shown in Figure 3.8 (which uses the so-called organization chart as the modelling vehicle; many of the rules and guidelines of classical administrative theory can be used in the design of such structures). Here, the key decisions fall under two headings, lower and upper structure. Lower structure decisions are concerned mainly with establishing the work roles of the supervisors and planners and the relationship between these people and the production supervisors. Initially, the lower structure is considered separately from the upper structure because it is influenced by — indeed, almost constrained by — the nature of the maintenance resource structure which, as explained, is in turn a function of the workload (see Figure 3.1).

The main task in establishing the upper structure is deciding on the relationship between the production group (the plant users), the maintenance supervisors and the trade groups (the maintenance doers), the maintenance planning group and the maintenance engineers (the technical decision makers). This complex problem is influenced by many factors, the most important of which are:

• the plant size, structure and geographical layout;
• the supervisory structure;
• the production administrative structure.

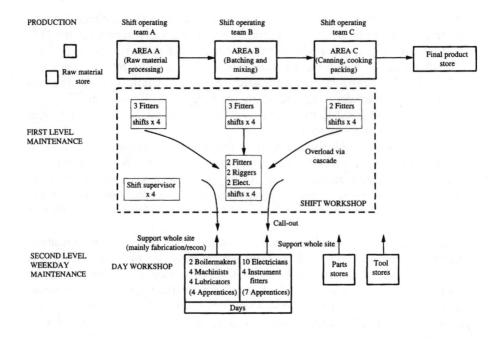

Figure 3.6. Weekday resource structure, food plant

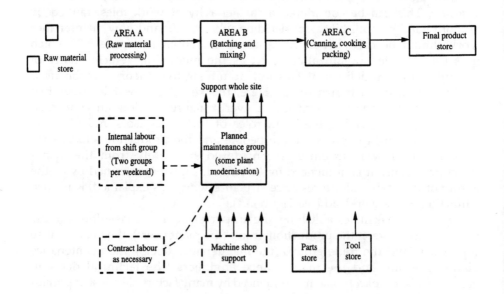

Figure 3.7. Weekend resource structure, food plant

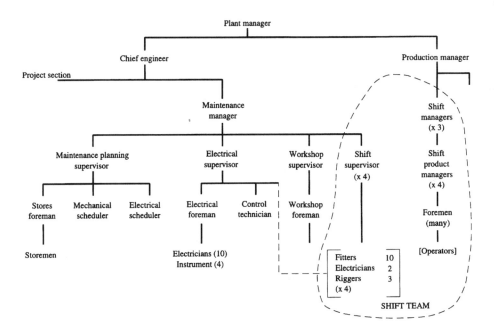

Figure 3.8. Administrative structure, food plant

The third element in the design of a maintenance organization is the formation of the maintenance systems, the most important of which is the *work planning system*, which defines the way in which the work is planned, scheduled, allocated and controlled. Figure 3.9 outlines such a system for the resource and administrative structure previously shown. The design of this should aim to get the right balance between the cost of planning the resources and the savings in the direct and indirect maintenance costs that result from use of such resources.

A *control system* is needed to ensure that the maintenance organization is achieving its objectives and to provide corrective action, e.g. change the life plan, if it is not. There would appear to be three principal vehicles for this:

(i) Control of maintenance *productivity*: ensuring that the budgeted levels of maintenance effort are being sustained and that required plant output is achieved.

(ii) Control of maintenance *effectiveness*: ensuring that the expected long-term and short-term plant reliability is being achieved, i.e. that the life plan is effective and is being carried out.

(iii) Control of maintenance *organizational efficiency*: monitoring the efficiency of utilization of workforce, materials and tools.

At this point it must be emphasized that the purpose of the organizational analysis that has been outlined is to enable the many decisions that affect the

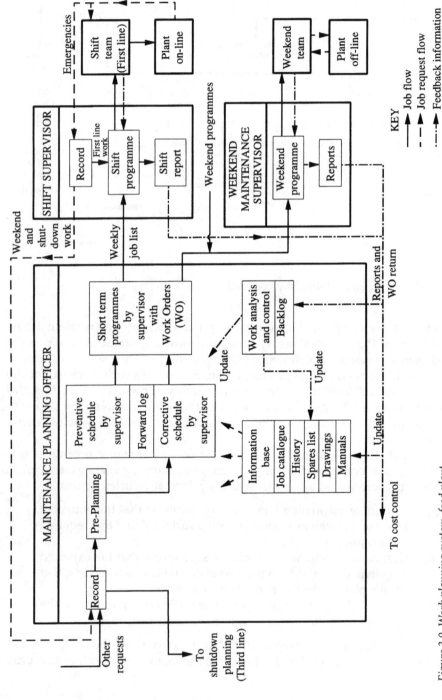

Figure 3.9. Work planning system, food plant

final shape of the organization to be seen in perspective. The possible effects of any particular decision on the complete organization and its objectives can then be better assessed. In the final analysis, the maintenance organization needs to be considered as a synergistic whole, i.e. an organism which is much more than merely the sum of its parts.

Figure 3.1 indicated that some form of formal *documentation* system — for the collection, storage, interrogation, analysis, and reporting of information (schedules, manuals, drawings or computer files) — is needed to facilitate the operation of all the elements of maintenance management. Figure 3.10, a general functional model of such a system (whether manual or computerized), indicates that it can be seen as comprising seven principal inter-related modules (performing different documentation functions). Considerable clerical

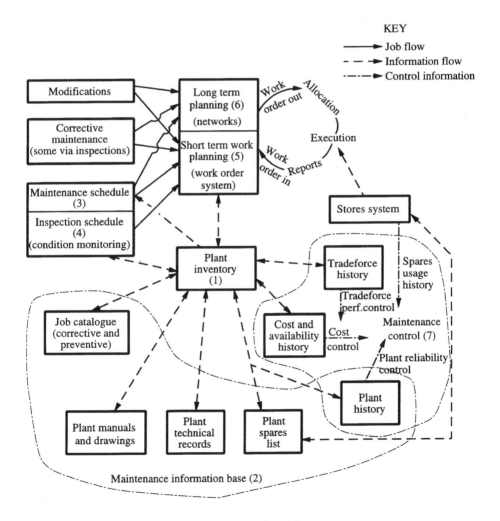

Figure 3.10. Maintenance documentation, general model

and engineering effort is needed to establish and maintain certain of these functions (e.g. the preventive maintenance information base). The control module, in particular, relies on an effective data collection system.

An analytical framework, based on various system models and which can facilitate understanding of a maintenance system — proposed or existing — has been outlined. Any maintenance organization will, however, be designed and operated by people and will have many interfaces with other parts of the company (notably with Production) which are also made up of people. It is therefore essential that this systems structure approach is complemented by a full consideration of the *human factors* in an organization.

Application of the approach

The organizational mapping that has been described was carried out in order to advise the particular food processing company about changing their maintenance strategy to meet a changing production demand. At the time of the first visit the plant's production pattern was three shifts per day, five days per week, fifty weeks per year. There was also considerable spare capacity; for example, only three lines out of four (see Figure 3.4) were needed to achieve full capacity. However, each line had its own product mix to satisfy the market demand. Thus, the availability of any given line for maintenance depended on the market demand and the level of finished product stored (which could be up to two weeks output). Off-line maintenance could therefore be carried out in the weekend windows of opportunity or, by exploiting spare capacity, during the week. In general, most of the off-line work was carried out during the weekend.

The problem the company faced was that they were going to increase capacity to 21 shifts per week. They wanted to know how this was going to affect their maintenance department. The author was asked to map their existing maintenance management systems and propose an alternative strategy which would facilitate 21-shift operation.

The effect of 21-shift operation on maintenance

The existing maintenance strategy was based on carrying out off-line main-tenance during the weekend windows and during the once-per-year holiday window. Little attempt had been made to exploit the excess capacity of the plant, or spare plant, to schedule off-line work while the plant was operating. The new 21-shift operating pattern meant that off-line maintenance would have to be carried out in this way. Indeed the strategy would have to move in the direction indicated in Table 3.2. This, in turn, would change the work-load pattern, i.e.

Table 3.2. Changes in maintenance strategy to accommodate new production policy

---

- A movement towards shutdowns of complete sections of plant based on the longest running time of critical units (e.g. the Hydros). The frequency of these shutdowns will, as far as possible, be based on running hours or cumulative output. However, for critical items, inspection and condition monitoring routines may be used to indicate the need for shutdowns, which will provide more flexibility and certainty about shutdown dates.

- All plant designated as non-critical, e.g. as a result of spare capacity, will continue to be scheduled at unit level (e.g. D line Filler/Closer).

- A much greater dependence on formalized inspections and condition monitoring routines, for the reasons given above and also to detect faults while they are still minor and before they become critical.

- A concerted effort either to design-out critical items or to extend their effective running time.

- the first-line workload would extend to 21 shifts;
- the off-line work (schedulable corrective and preventive) would need to be done during the weekday day shifts;
- the third-line major work could still be carried out in the holiday window.

Thus, to cover this workload, the maintenance organization would also have to change. The most likely organization would be based on a first-line, 21-shift, group (perhaps with a reduction in manning-per-shift) and a second-line day group operating five days per week. This, in turn, would influence the administrative structure and work planning systems. The latter would have to improve considerably.

The strategic thought process

The example shows that the maintenance department requires managerial strategic analysis in the same way as any other industrial department. The thought process that was involved is indicated in Figure 3.11. The process starts with the Sales–Production reaction to market demand, the resulting change in the plant operating pattern and the increased plant operation time. This, in turn, requires amended maintenance life plans and a modified maintenance schedule. Thus, the maintenance workload changes, which brings in train the

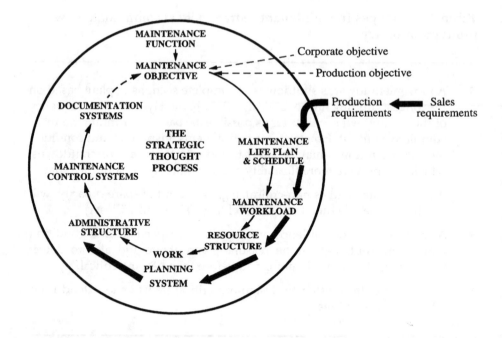

Figure 3.11. The influence of market demand on maintenance strategy

need to modify the maintenance organization and systems. Understanding and applying this type of strategic thought process is the cornerstone of effective and fruitful maintenance management analysis.

4
The structure of plant

Introduction

The main purpose of this book is to present and discuss methods for analysing the complex problem of setting up maintenance strategies for industrial plant that are both effective and efficient. In Chapter 3 the various necessary elements of a maintenance system were described, and illustrated by reference to the particular case of a food processing plant. This made it clear that the central problem is indeed the formulation of strategy. However, before enlarging on this we need to discuss:

(a) methods of modelling and analysing the operation of industrial plant, in ways that shed light on maintenance strategy;
(b) the reasons why such maintenance is needed;
(c) what exactly is meant by maintenance *strategy*.

Modelling industrial plant

An industrial plant could be anything from a food processing plant, as in Chapter 3, to an alumina refinery or a batch chemical plant. It will be required to carry out some overall production function, usually for some anticipated life. The effectiveness of the production function is usually measured in terms of its rate of output (e.g. tonnes of alumina per week).

In order to discuss an industrial plant in terms of its maintenance strategy it is useful to create an analytical model of its structural and process flow characteristics (and also of its reliability dependencies, a topic that will be introduced in Chapter 5). One way of regarding it is as a hierachy of parts, ranked according to their functional dependencies into units, assemblies, sub-assemblies and components (see Figure 4.1). A unit — a reaction vessel in a chemical plant (see Figure 4.2), for example, or one of the mixers in the food processing plant — can be defined as 'a collection of items inter-connected mechanically and/or electrically to perform a specific production sub-function of the plant'. A key part of this definition is the recognition that the unit performs a production function (e.g. generates so many tonnes of milled bauxite per hour).

Most industrial plant can be informatively represented, or modelled, by a diagram showing the process flow between its various units. In this way Figure 4.3, for example, charts the overall operation of a batch chemical plant

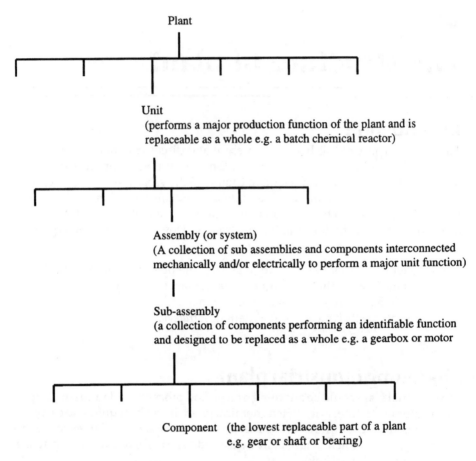

Figure 4.1. *A plant modelled as a hierarchy of parts*

and Figure 4.4 the detailed flows of one of its reaction streams. Models of this kind are an essential aid to understanding the production characteristics of plant and hence the cost, safety and scheduling considerations which have to be taken into account when determining maintenance strategy.

Each of the units can itself be informatively sub-divided into a hierarchy of parts ranked largely according to their replaceability. In Figure 4.5, for example, the reaction vessel of Figure 4.2 is analysed into assemblies such as the agitator drive, sub-assemblies such as the drive motor, and finally into components — the lowest level of replaceability — such as gears. This kind of analysis is particularly important when setting up the equipment inventory (usually based around the unit) and is especially useful when identifying the maintenance-causing assemblies, sub-assemblies and components of the unit.

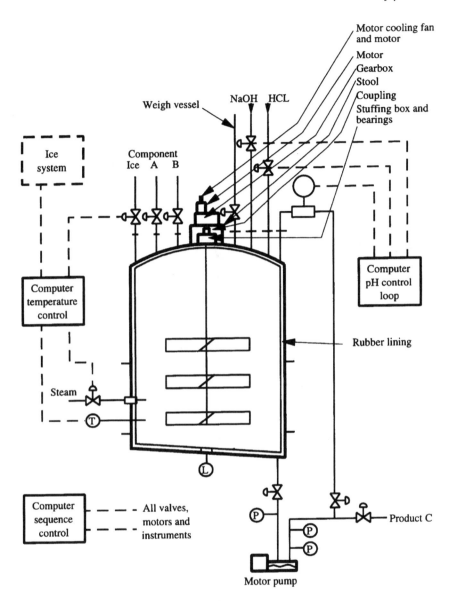

REACTOR FUNCTION:

> To perform a defined stage of a production process
> i.e. chemical reaction under controlled pH and temperature

Figure 4.2. A batch chemical reactor

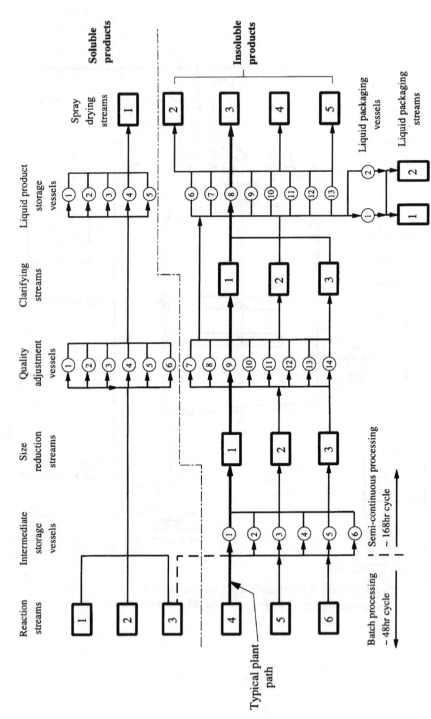

Figure 4.3. Process flow diagram, batch chemical plant

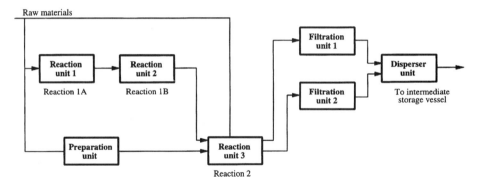

Figure 4.4. Process flow diagram, a reaction stream

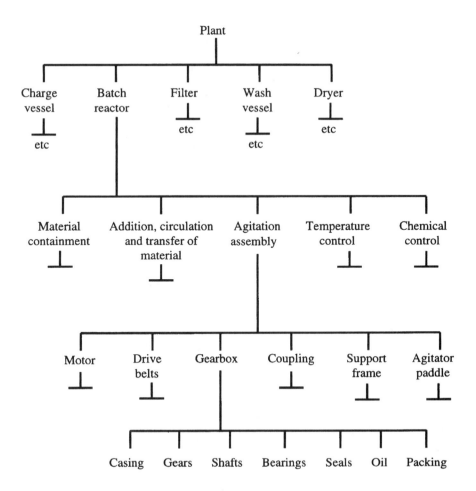

Figure 4.5. Hierarchical division of a batch chemical plant

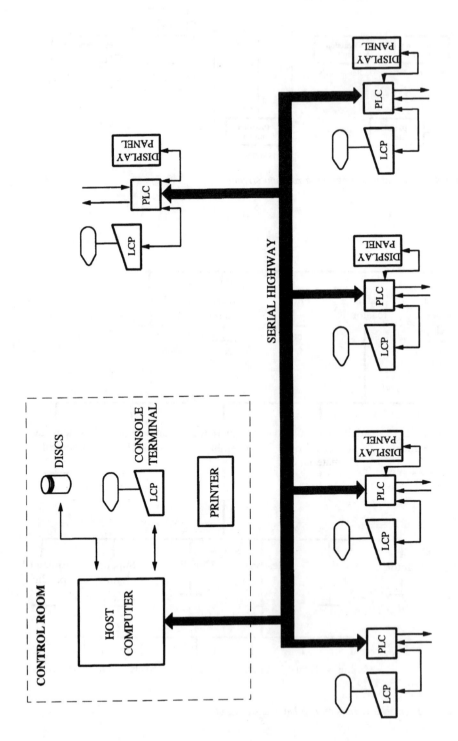

Figure 4.6. Control system for a chemical plant

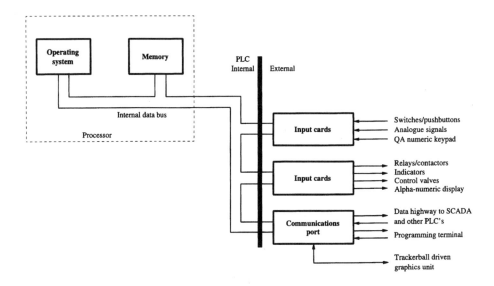

Figure 4.7. PLC internal/external block diagram

A similar analysis can be carried out of the plant control systems (see Figures 4.6, 4.7 and 4.8)

The reason for maintenance

The need for maintenance originates at component level (e.g. at the bearings itemized in Figure 4.5). When a component is unable, according to some pre-determined criterion, to perform its designated function it can be said to have failed, and this could be a *complete* or a *partial* loss of function. Such a loss could be contained at unit level (temporarily, at least) or have consequences at plant level, depending on the design of the plant, e.g. on the availability and capacity of inter-stage storage or redundancy (hence the importance of modelling plants in the way shown in Figures 4.3 and 4.4). The loss of function could also have safety or product quality consequences.

For technological and economic reasons, many of the components of a plant will have been designed to have a useful life greater than the longest plant production cycle but less than that of the plant itself. In most cases such maintenance-causing parts, especially the short-life ones, will have been identified at the design stage and made easily replaceable at component level.

Other components will fail for reasons that are not easy to anticipate — such as poor design, poor maintenance, or malpractice — and may be extremely expensive to replace, often requiring a substitution at a higher level of plant, i.e. of the complete assembly. In addition, as the plant ages, failure rates

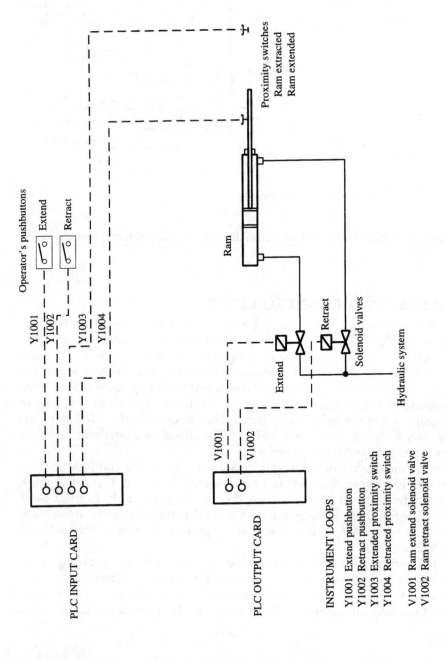

Figure 4.8. Local control system showing six loops

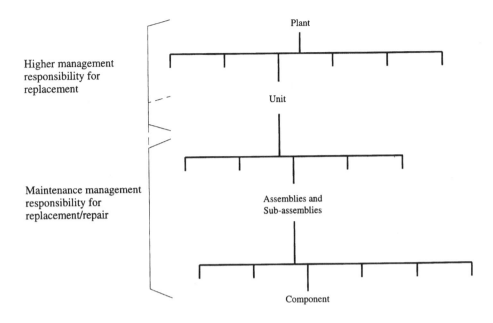

Figure 4.9. Levels of managerial responsibility for replacement/repair decisions

and maintenance replacement costs can be expected to increase as the long life, expensive, components—and eventually the assemblies and whole units—reach the limits of their useful lives.

Capital replacement policy

The delegation of the responsibility for taking replacement and repair decisions differs from one organization to another, but usually it is higher management that has to decide such matters for major units (or, indeed, for the complete plant) and maintenance management for assemblies and below (see Figure 4.9). This division of responsibility is obligatory because the policy for major units and above is influenced by external, often longer term, factors such as obsolescence, sales trends, or movements in the cost of capital as well as internal, shorter term, factors such as operating and maintenance costs. Decisions involving the replacement of complete units or sections of plant should really be regarded as part of capital replacement policy.

Maintenance strategy

Maintenance can be considered as the replacement or repair of components and assemblies (before or after failure) so that the unit concerned can perform its

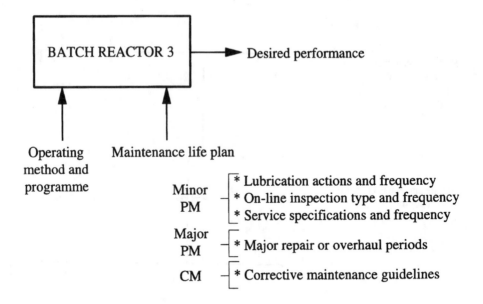

Figure 4.10. A typical unit and its maintenance life plan

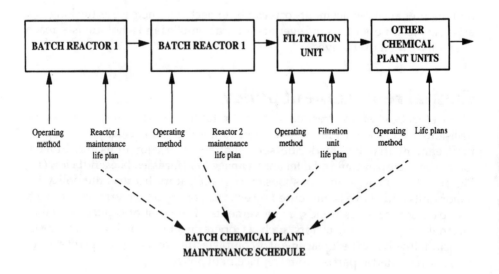

Figure 4.11. Assembling a maintenance schedule

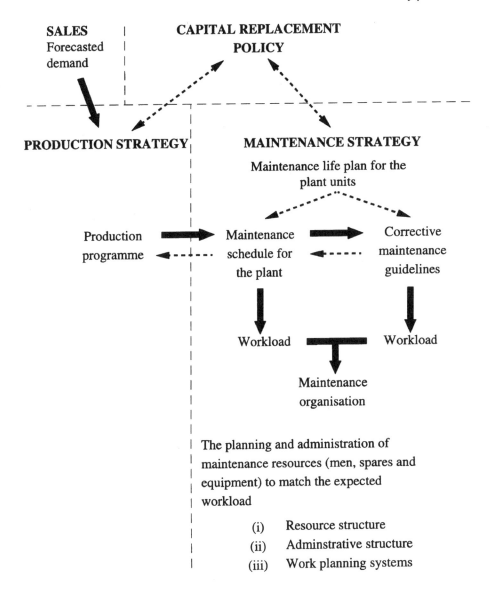

Figure 4.12. Relationship between maintenance strategy and capital replacement policy

designated function over its expected life.

A maintenance strategy involves the identification, resourcing and execution of many thousands of repair, replace and inspect decisions. It is concerned with:

- formulating the *best life plan* (see Figure 4.10) for each unit. This is a comprehensive programme of maintenance procedures — repair/ replace/inspect at various frequencies — spanning the expected life of the unit;

- formulating a *maintenance schedule* for the plant (see Figure 4.11). This should be assembled from the programmes of work contained in the unit life plan(s) but should be dynamic, e.g. readily adjustable in the light of changes in the production schedule;
- Establishing the organization to enable the scheduled, and other, maintenance work to be resourced (see Figure 4.12, which also shows that maintenance strategy and capital replacement policy are interrelated, i.e. maintenance cost influences unit replacement decisions and vice versa).

5
The reliability of plant components*

Introduction

The main thrust of the discussion in Chapter 2 was that the development of long-term strategies for achieving effective and efficient industrial maintenance management should be rooted in the life-cycle ideas of the Capital Asset approach. It was explained that, for their realization, these ideas require, among other things, the specification, assessment, monitoring and control of plant *reliability*.

Reliability was continued as a basic theme into Chapters 3 and 4, it being stated in the former, that ensuring and controlling it is the basic function of a maintenance department, and in the latter, that to formulate an appropriate strategy for carrying out that function the desirable first step is to create analytical models of the plant concerned — models not only of its structural and flow characteristics but also of its reliability dependencies (the relationships, often quite complex, between the plant's overall availability, say, and that of the various units of which it is composed). So far, however, what exactly is meant by the term 'reliability' has been taken for granted; it has been assumed that the reader will have, at the very least, a general idea of what it means, based either on experience or intuition. To pursue the topic further we now need to define the term precisely and to examine how, in the case of both engineering plant items and plant systems, it may be quantified, measured and predicted.

In this chapter the focus will be on the reliability of individual engineering components or replaceable sub-assemblies (e.g. pumps, valves, etc., see Figure 4.1). It will be shown how the sort of statistical data on component failure that can be collected via the maintenance function can be analysed to give quantitative measures of reliability, and information on patterns of reliability behaviour (or modes of failure), that are of great value in maintenance decision-making. (Chapter 6 will explain how such data can then be combined to produce assessments of the reliability or availability of complete systems.) Appreciation of the reliability concepts introduced here is an essential pre-requisite to understanding the ideas of strategy formulation that will be developed later, in particular in Chapter 8 (Principles of preventive

* Chapter contributed by M. J. Harris, Honorary Fellow, University of Manchester School of Engineering.

maintenance), Chapter 9 (The top-down bottom-up approach) and Chapter 13 (Reliability Centred Maintenance).

Engineering reliability, probability and statistics

The solution of a maintenance problem often requires us to think about the reliability of a particular equipment. For example, we might need to assess whether an especially critical compressor is likely to run satisfactorily, without failing, from one seasonal shutdown to another. (If not, should it be replaced, maintained differently, or should we acquire a standby?) Now we know that we can never be *absolutely* certain that it will. Even if it is well designed, well maintained and carefully operated there is always a small chance that it will fail in service, and we cannot say exactly when this might happen.

Thus, analysis of engineering reliability — like betting on the horses — deals in *probabilities* (i.e. likelihoods of success or failure) and *probabilistic variables* (i.e. quantities that vary randomly, such as times to failure). In particular, it deals with the application of statistical techniques for analysing patterns of failure of components and equipment. Now probability and statistics are, in their own right, major mathematical (and philosophical) topics, in certain areas still very contentious, and ones to which many large and profound tomes have been devoted. However, the reader will be comforted to know that we shall not be going in to them in any great depth. Indeed, it is the author's experience that most of the probabilistic aspects of reliability analysis that need to be coped with for engineering purposes can be understood fairly intuitively, or anyway via simple comparisons (e.g. with dice throwing, or with the aforementioned betting on horses!).

Item reliability

A good definition of this was given by Green and Bourne in their authoritative textbook[1]:

> The probability that an item will perform its required function in the desired manner under all the relevant conditions and on the occasions, or during the time intervals, when it is required so to perform.

Like most definitions, some of the words need clarification. Let us take them in turn. First, *probability*. This is still a subject of philosophical debate. For our purposes, however, we will take it to be a measure of what is expected to happen, *on the average*, if a given event is repeated a large number of times under identical conditions, e.g. the probability of getting a five, say, when throwing a six-sided die is 1/6, or 16.67 per cent.

Next, an *item*. As used by Greene and Bourne (NB in Chapter 8, 'item' is given a very specific meaning, referring to *replaceability*, and is used in that particular sense thereafter) this could be:

a component:	the smallest part which would be replaced or repaired on failure (e.g. a drive belt), or a directly replaceable sub-assembly (e.g. an electric motor);
a unit:	comprising a number of components and sub-assemblies (e.g. a batch reactor);
a plant:	comprising many units (e.g. a complete reaction stream).

Reliability analysis is quite different in each case and it is therefore important to clearly define the physical boundaries of the systems* being analysed.

The phrase '... its required function in the desired manner under all the relevant conditions ...' refers to the *duty* undertaken. With electronic equipment the duty demanded of any particular component will probably be much the same in one application as in another; stresses will usually be low and steady, and the equipment encapsulated. The reliability observed in one context is therefore likely to be similar to that in another, i.e. to be *generic* or 'characteristic of a large group or class; general, not specific or special' (to quote from the Oxford dictionary). This would rarely be the case with mechanical or hydraulic equipment, which could be subject to wide ranges of operating stress (start-up acceleration, throttled or open running, etc.) and to environmental extremes (tropic or arctic, off or on shore, etc.), so reliability assessment would have to take this into account.

The phrase ... *on the occasions, or during the time intervals* ... indicates that there are two basic sorts of reliability, namely:

(a) *Time independent*
The component functions only on demand, being otherwise dormant (e.g. a pressure release valve). Reliability is measured by the *probability of successful function*, P_S, e.g. if a starter motor has failed to operate three times in a hundred demands, say, then

$$P_S = 97/100 = 0.97 \text{ or } 97\%$$

It is more meaningful, and therefore it is the customary practice, to quote this in terms of the *probability of failure on demand* (or *fractional dead time, FDT*), P_F, i.e.

$$P_F = 3/100 = 0.03 \text{ or } 3\%$$

(b) *Time dependent*
The component functions continuously (e.g. an electric motor). Reliability is measured by the probability $R(t)$ that it will run successfully for some

---

* For the purposes of this book a 'system' is taken to mean either an inter-connected set of components or sub-assemblies performing an identifiable unit function (e.g. the agitator system of Figure 4.5) or an inter-connected set of units performing an identifiable plant function (e.g. the reaction stream of Figure 4.4).

specified time t (e.g. from one annual shutdown to the next). Thus, if a hundred identical pumps had been started together and after three weeks twenty, say, had failed then

$$R(3 \text{ weeks}) = 80/100 = 0.8 \text{ or } 80\%$$

In this chapter we will focus on reliability of type (b), because it is of relevance to the maintenance of practically all types of *continuously operated* industrial plant. Of course, if the main concern was the maintenance of, say, *safety* or *standby* equipment, e.g. smoke alarms, then reliability of type (a) would be of primary interest. A concise, but useful introduction to the assessment of reliability of this latter kind is given in the UK Institution of Mechanical Engineers Process Industries Guide, *The Reliability of Mechanical Systems*[2].

Statistical analysis of component lifetimes

Mean, variance and standard deviation

Let us assume, as a highly idealized illustration, that we have been able to test 100 identical pumps of a new design by running them continuously until each one has failed, with the results shown in Table 5.1.

Table 5.1. Pump failure data

Time to failure (hours)	No. of pumps failing	Fraction failing	Fraction failing per hour
Class interval	*Frequency*	*Relative frequency*	*Relative frequency density*
300–399	2	0.02	0.0002
400–499	9	0.09	0.0009
500–599	21	0.21	0.0021
600–699	40	0.40	0.0040
700–799	19	0.19	0.0019
800–899	8	0.08	0.0008
900–1000	1	0.01	0.0001
Totals	100	1.00	

Note: (a) the second row of the table shows the standard statistical terms ('Class interval' etc.) for the types of quantity evaluated; (b) the figures in the fourth column are obtained by dividing those in the third by 100 hours, the width of the class interval used.

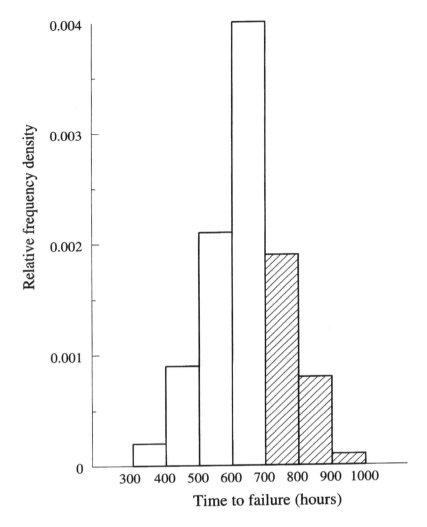

Figure 5.1. Histogram of pump failure data

Using the data in the fourth column a *histogram* can be constructed, as in Figure 5.1. (By drawing it this way, i.e. with the y-axis representing the *probability density*, the area of the block above each class interval always equals the *relative frequency* of failure in that time interval, even if unequal class intervals were to be used, which is sometimes more convenient.)

The assumption might now be made that the pattern of failure of this sample is typical of all such pumps; i.e. the *observed* relative frequencies truly reflect the *expected* probabilities of failure. The probability that any one pump of this kind will last longer than, say, 700 hours is then given by the shaded area in the histogram, i.e.

$$0.19 + 0.08 + 0.01 = 0.28 \text{ or } 28\%$$

We now require some numbers which will indicate the *general nature* of the variable quantity (or of the *variate* as it is called in statistical terminology) that we have observed.

(i)   For its average magnitude, or *central tendency*, we use the *arithmetic mean*

$$m = (0.02 \times 350) + (0.09 \times 450) + (0.21 \times 550) + \ldots\ldots \text{ etc.} = 642 \text{ h}$$

where, for example, in the first bracket 0.02 is the relative frequency and 350 hours the mid-point, or *class mark*, of the first quoted class interval.

(ii) For the spread, or *dispersion*, we shall calculate the *variance*

$$s^2 = 0.02(350 - 642)^2 + 0.09(450 - 642)^2 + \ldots\ldots\ldots \text{ etc.} = 13\,500 \text{ h}^2$$

where, as before, the first bracket, say, refers to the data for the first quoted class interval and 642 hours is the previously calculated overall mean. A quantity measured in hours-squared is rather mysterious (although it is, in fact, indispensable in most statistical calculations), so for presenting information on the observed spread of the times-to-failure we quote its square-root, the *standard deviation*

$$s = (13\,500)^{1/2} = 116 \text{ hours}$$

Probability density functions

If many thousands of pumps had been tested, instead of just one hundred, the width of the class intervals in Figure 5.1 could have been reduced and a virtually continuous *probability density function* or pdf obtained, as in Figure 5.2. Many failure processes generate pdfs of time-to-failure which can be represented fairly accurately by simple mathematical expressions. This can be useful in reliability calculations.

The Normal or 'wear-out' pdf

Many engineering items exhibit definite wear-out, i.e. they mostly fail around some mean operating age, although a few fail sooner and a few later. The distribution of times-to-failure often approximates to the symmetric, bell-shaped, *Normal* pdf, a distribution which is of pivotal importance in statistical theory. (It is often called *Gauss's* distribution because he derived it — by formulating a simple model of the way in which errors of measurement are generated.) If the times-to-failure were to be distributed in this way then 50 per cent of them would fall in the range $(m - 0.67s)$ to $(m + 0.67s)$, and 95 per cent

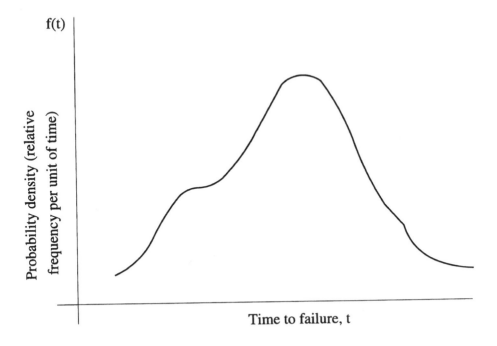

Figure 5.2. Continuous probability density distribution

in the range $(m - 2s)$ to $(m + 2s)$, where m is the measured mean of the distribution and s its standard deviation (see Figure 5.3). Statistical tables[3] give other percentage probabilities for other ranges (expressed as multiples of s) about the mean.

The negative exponential, or 'random failure' pdf

During their 'as-designed' lives many engineering components, if properly operated, do *not* 'wear-out'. On the contrary, they are as likely to fail sooner as later. The probability of failure is constant (and probably small), i.e. the component is always effectively 'as good as new'. This indicates that the cause of any failure is external to the component, e.g. overload. It can be shown that, in this case, the distribution of time-to-failure t is given by the *negative exponential pdf* (see Figure 5.4), i.e.

$$f(t) = \lambda \exp(-\lambda t)$$

where λ = mean failure rate (failures/unit time/machine) and
$1/\lambda$ = mean time to failure (MTTF).

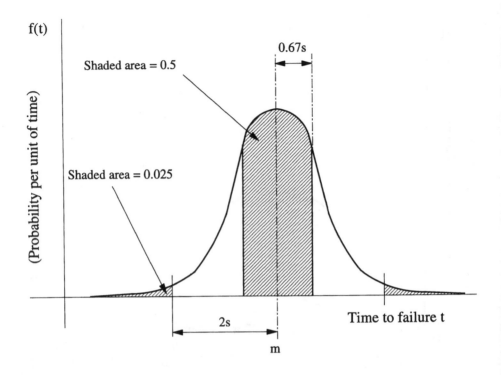

Figure 5.3. The Normal probability density function

The hyper-exponential, or 'running-in' pdf

Sometimes, the probability of failure is found to be higher immediately after installation or overhaul than during subsequent operation. This can be represented by a pdf of time-to-failure which exhibits two phases, an initial rapid fall and a later slower one (see Figure 5.4). This is evidence that some of the components concerned have manufacturing defects, or have been re-assembled incorrectly, faults that show up during the running-in period. Components that survive this period are without such defects and go on to exhibit the sort of time-independent failure probability previously discussed. Note — they are not improving with age! Some components merely start off with a better chance of survival than others.

Measures of component reliability

The information displayed in Figure 5.1 can be presented in other ways that may be more useful, notably as plots of:

- cumulative, or failure, probability function, $F(t)$;
- reliability function, $R(t)$;
- hazard rate function, $Z(t)$.

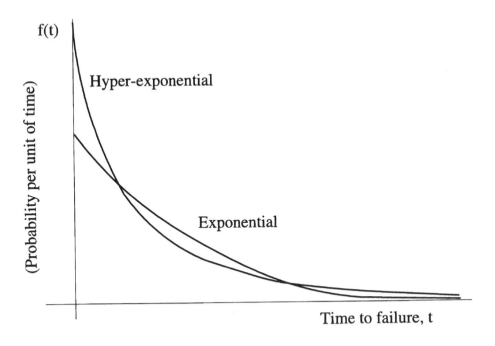

Figure 5.4. *Hyper-exponential and exponential pdfs*

Table 5.2. Rise in total fraction failed: data of Table 5.1

Time (h)	t	0	100	200	300	400	500	600	700	800	900	1000
Total fraction failed	$F(t)$	0	0	0	0	0.02	0.11	0.32	0.72	0.91	0.99	1

For example at $t = 600$ h, $F(t) = 0.02 + 0.09 + 0.21 = 0.32$ (see third column of Table 5.1)

Failure probability F(t)

A graph can be drawn (see Table 5.2 and Figure 5.5) of the rise in the total fraction $F(t)$ of pumps failed by a given time t.

If the tested pumps are representative, the observed value of $F(t)$ at any particular running time t can be taken to be the probability, for *any* such pump, of failure before that time. The mathematical expression of a plot such as that in Figure 3.5 is called a *cumulative distribution function* or cdf. For the negative exponential pdf, for example, the cdf is

$$F(t) = 1 - \exp(-\lambda t)$$

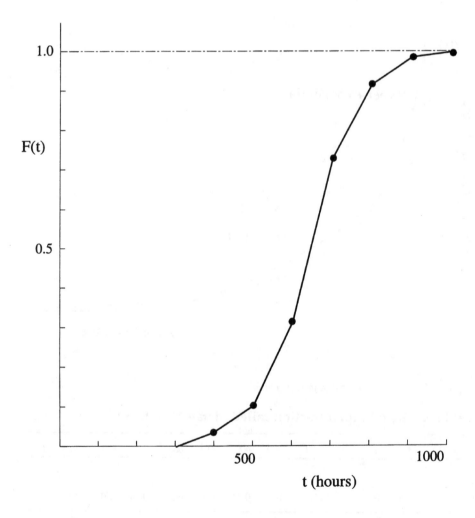

Figure 5.5. Probability, F(t), of failure before time t: pump data

Reliability, R(t)

Table 5.2 showed the fraction $F(t)$ of components *failed* before running time t. Alternatively, we could tabulate and plot, as in Table 5.3 and Figure 5.6, the fraction $R(t)$ *surviving*, i.e. the measure of *time-dependent reliability* discussed earlier. In most of the literature on this subject it has now become the custom to call this the *reliability*, without qualification. Clearly, $R(t) = 1 - F(t)$. So, for the negative exponential case, for example,

$$R(t) = \exp(-\lambda t)$$

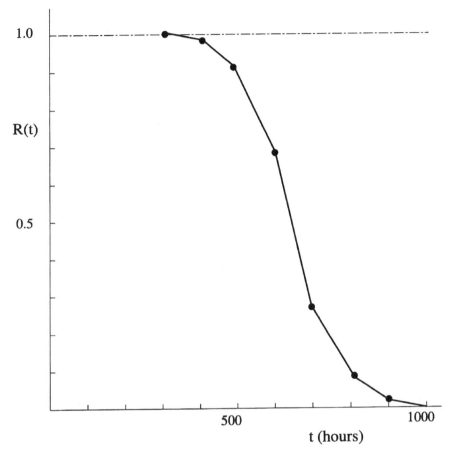

Figure 5.6. Reliability, R(t), at time t: pump data

Again, if the pumps are representative $R(t)$ can be taken to be the estimated *reliability* at time t for any such pump.

Table 5.3. Fall in total fraction surviving: data of Table 5.1

Time (hours)	t	200	300	400	500	600	700	800	900	1000
Fraction surviving	$R(t)$	1	1	0.98	0.89	0.68	0.28	0.09	0.01	0

Hazard rate, Z(t)

This is defined as:

> the fraction, of those components which *have* survived up to the time t, expected to fail, per unit time

Thus, at any time t,

$$Z(t) \;=\; \frac{\text{fraction of original pumps failing per hour at time } t}{\text{fraction of original pumps still running at time } t}$$

$$=\; \frac{f(t)}{R(t)}$$

So, for the negative exponential case,

$$Z(t) \;=\; \frac{\lambda e^{-\lambda t}}{e^{-\lambda t}} = \lambda$$

i.e. the failure rate is constant, the component is always 'as good as new', as already explained. For the data of Table 5.1, $Z(t)$ is calculated, tabulated and plotted in Table 5.4 and Figure 5.7.

Table 5.4. Variation of hazard rate: pump data of Table 5.1

Time (hours) t		350	450	550	650	750	850	950
Fraction, of original pumps, failing per hour, at time t	$f(t)$	0.0002	0.0009	0.0021	0.0040	0.0019	0.0008	0.0001
Fraction surviving at time t (estimated from Figure 5.6)	$R(t)$	0.99	0.94	0.79	0.48	0.19	0.05	0.005
Hazard rate (conditional failure probability per hour)	$Z(t)$ $= f(t)/R(t)0$	0	0	0.01	0.01	0.0160	0.02

In Figure 5.8, $F(t)$, $R(t)$ and $Z(t)$ are compared for the three basic types of failure — running-in, useful-life and wear-out.

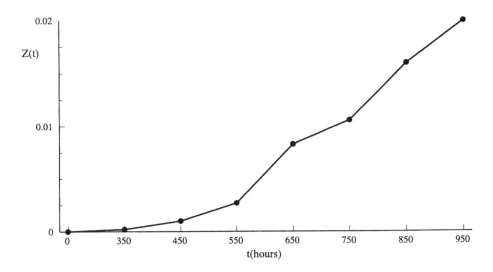

Figure 5.7. Hazard rate plot for pump data

Time-to-failure

Our frequent use of the expression 'time-to-failure' needs to be qualified. *Time* may, of course, mean *actual calendar time* (before failure). In many contexts, however, it might more appropriately refer to *total running time*, i.e. exclusive of stoppages, or *number of operational cycles*.

What constitutes a *failure* depends on the operational requirement, even for the simplest electrical component. Failure of a resistor could be a break, or a short, or either one of these. There may be several different ways in which the operation of a mechanical component could be degraded, but only one or two which would actually impair *unit* performance should they occur, e.g. a hydraulic valve could suffer an internal leak, an external leak, failure to close, failure to open, or spurious operation; also, a partial leak may be acceptable.

The whole-life item failure profile

By combining the three $Z(t)$ curves of Figure 5.8 a single $Z(t)$ curve as in Figure 5.9 can be obtained which, broadly speaking, gives the whole-life profile of failure probability for the generality of components. This is the much quoted — and much abused — 'bath-tub curve'.

This is only the 'bath-tub curve' when the variable on the y-axis is the hazard rate, $Z(t)$, as we have defined it here. The actual level of $Z(t)$, the time scale involved and the relative lengths of the three phases, will vary by orders of magnitude from one sort of component, and one application, to another.

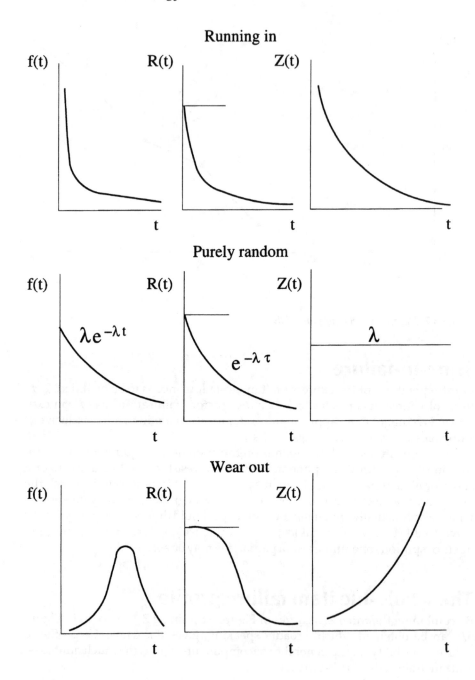

Figure 5.8. Principal modes of failure

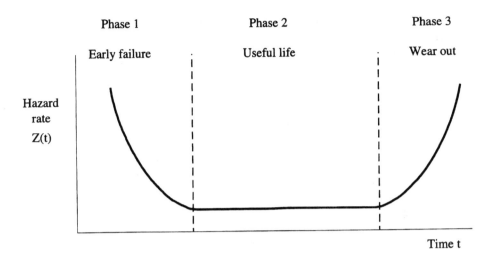

Figure 5.9. Typical Z(t) characteristic for engineering devices

Furthermore, in any specific case one or two of the phases could be effectively absent (e.g. in the case of high-reliability aircraft control gear, where running-in failure is negligible and wear-out non-existent). Human beings strikingly exemplify this behaviour. Figure 5.10 is taken from Greene and Bourne[1]. It gives the hazard rate (where the hazard is death, and the hazard rate is called, by actuaries, the *force of mortality*) for the UK male population in the 1960s.

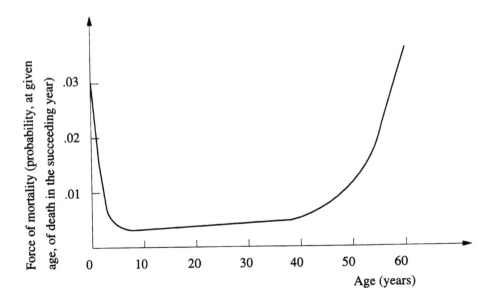

Figure 5.10. Death rate characteristic for males living in England and Wales for the years 1960–62

Examination of Figure 5.9 suggests that if a component's whole-life behaviour conformed to that shown, and the various hazard rates and time scales were known, then perhaps we should:

(a)   run all such components in, for the duration of Phase I, before putting into service;

(b)   carry out nothing other than minimally intrusive on-line routines, such as lubrication and simple condition-monitoring, during Phase II (the 'useful life');

(c)   replace or overhaul them early in Phase III.

In practice, of course, the statistical information would rarely be known, and even if it were there would be many other factors — output needs, economics, opportunities — to take into account when identifying appropriate maintenance policy, as will be discussed in depth in Chapters 8 and 9.

Diagnosis of recurrent failures and prescription of the remedy

As has been shown, the shape of the pdf of time-to-failure or, more strikingly, the trend of the hazard rate plot can indicate whether a recurring failure is of the running-in, wearing-out or purely random kind. Combined with appropriate physical or chemical investigations (e.g. microscopic examination of wear or fracture surfaces, chemical analysis of lubricants) this information can facilitate the diagnosis of the cause of the failure and prescription of remedial action. If, for example, the failure was clearly a wear-out the choice would be between time-based, condition-based or failure-based replacement, or design-out. The final judgement, however, would probably have to take into account a combination of economic, environmental and safety factors, as will be fully discussed in Chapter 8.

Weibull analysis of item lifetimes

Although, as explained, the Normal pdf can be used to represent wear-out failure, the exponential pdf purely random failure, and the hyper-exponential pdf running-in failure (and statisticians have, in fact, conjured up many other such pdfs for these and similar purposes), there is one particular pdf, the *Weibull* pdf , which, because it can represent *any* of the three basic types of failure, has been found to be particularly useful. In addition, it has two other sovereign virtues:

(i)   it can be applied via simple graphical techniques;

(ii)   it is expressed by a formula in which all the terms have engineering significance.

The ideas underlying this pdf may be grasped from Weibull's own derivation,

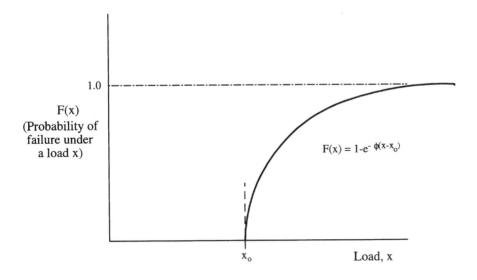

Figure 5.11. Probability of failure of specimens as a function of load

which was neither mathematical nor statistical, but was based on a few practical considerations.

Weibull was involved in analysing the results of load tests on many nominally identical test specimens of a particular type of steel. Their ultimate tensile strengths exhibited random variability, as they always do. If $F(x)$ was defined as the cumulative fraction which exhibited strengths *less* than a particular load x (i.e. $F(x)$ was the *cumulative distribution function* or cdf, the distribution of the probability that a specimen would fail under the load x), then a plot of $F(x)$ looked like the one shown in Figure 5.11. None failed before some given load x_0 (the *guaranteed* strength) and a very few hung on to quite large loads.

First, Weibull conjectured that it might be possible to represent such a cdf fairly accurately by the expression

$$F(x) = 1 - \exp\{-\phi(x - x_0)\},$$

where $\phi(x - x_0)$ would be some function of $(x - x_0)$, as yet undefined and which itself increased as x increased, e.g. $3(x - x_0)$, or $(x - x_0)^2$, or whatever. This would give a plot which started at x_0 and approached $F(x) = 1$ asymptotically, as required. However, $\phi(x - x_0)$ would have to be such that it gave the appropriate rate of rise of the value of $F(x)$, and would have to be dimensionless (because it is an exponent, a power).

Weibull found that the form

$$\phi(x) = \left\{\frac{x - x_0}{\eta}\right\}^\beta$$

where η was a characteristic load (determining, along with x_0, the scale of the loads involved) and β was a curve-shaping factor, gave him an expression for the cdf,

$$F(x) = 1 - \exp\left\{-\left(\frac{x - x_0}{\eta}\right)^{\beta}\right\}$$

which enabled him to correlate his test data very well. In addition, the expression had some other very useful properties, as we shall see.

In the reliability problems that we are looking at here the stressing factor is not load but *running time t*, since new or since last overhaul. The Weibull cdf for *times to failure* is therefore written as:

$$F(t) = 1 - \exp\left\{-\left(\frac{t - t_0}{\eta}\right)^{\beta}\right\}$$

From this, some not very complicated mathematics (for example, $R(t) = 1 - F(t)$) then leads to the appropriate expressions for the Weibull pdf $f(t)$, reliability $R(t)$, and hazard rate $Z(t)$:

$$f(t) = \frac{\beta(t - t_0)^{\beta - 1}}{\eta^{\beta}} \exp\left\{-\left(\frac{t - t_0}{\eta}\right)^{\beta}\right\}$$

$$R(t) = \exp\left\{-\left(\frac{t - t_0}{\eta}\right)^{\beta}\right\}$$

$$Z(t) = \frac{\beta}{\eta^{\beta}}(t - t_0)^{\beta - 1}$$

Each of the constants in these formulae has a practical meaning and significance.

The *threshold time-to-failure*, or *guaranteed life* t_0. In many cases of wear-out the first failure does not appear until some significant running time t_0 has elapsed. In the Weibull expressions the time factor is then the time interval $(t - t_0)$.

The *characteristic life*, η. When $t - t_0 = \eta$, $R(t) = \exp(-1) = 0.37$, i.e. η is the interval between t_0 and the time at which it can be expected that 37 per cent of the items will have survived (and hence 63 per cent will have failed).

The *shape factor*, β. Figure 5.12 shows how the Weibull pdf of time-to-failure changes as β is changed (for clarity, on each plot $t_0 = 0$ and $\eta = 1$).

If β is significantly less than one the pdf approximates to the hyper-exponential, i.e. is characteristic of 'running-in' failure.
If $\beta = 1$ the pdf becomes the simple negative exponential, characteristic of 'purely random' failure.

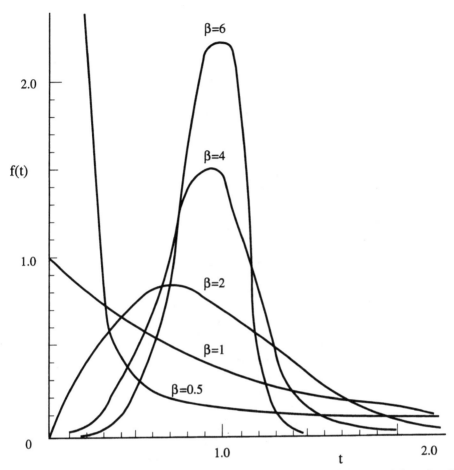

Figure 5.12. Influence of shape factor β on the form of the Weibull pdf of time to failure (for all plots $t_0 = 0$ and $\eta = 1$)

As β rises above a value of about 2 the pdf converges ever more closely to the Normal pdf characteristic of 'wear-out' failure.

NB For the first two cases t_0 must be zero, of course; for the wear-out case it may or may not be. Also note from Figure 5.12 that β characterizes the consistency of failure occurrence. The larger its value the greater is the tendency for the failures to occur at about the same running time.

Weibull probability paper

How do we test whether several times-to-failure, collected from the history record of a particular type of component, look as if they could be plausibly represented by a Weibull cdf? In the language of statistics, whether they look as if they have been sampled from such a distribution? And if they do, how do we determine the values of t_0, η and β which will give the distribution which best fits the data?

One easy way is to use *Weibull probability graph paper*. There are several versions of this; we shall use the one that is probably the most widely used in the UK, marketed by the Chartwell technical graph paper company (Ref No. 6572 in their list. A full size blank copy is printed in the IMechE mechanical reliability guidebook[2]) On this, the y-axis variable is the cumulative fraction failed, $F(t)$, expressed in per cent, and the x-axis variable is the $(t — t_0)$, in whatever are the appropriate units of time for the particular component studied (as explained earlier in this section, 'time' in this context is a measure of usage and might appropriately be 'number of operational cycles'). The axial scales are so arranged that if a *theoretical* Weibull cdf were to be plotted on the paper (i.e. using values of $F(t)$ calculated from the expression given earlier) they would lie on a perfectly straight line. The following example shows how the paper is used.

A Weibull analysis of a large and complete sample of times-to-failure

One hundred identical pumps have been run continuously and their times-to-failure recorded. To fit a Weibull expression to the data we proceed as follows.

1. The data are tabulated as in Columns 1 and 2 of Table 5.5.
2. Successive addition of the figures in Column 2 leads to Column 3, the *cumulative* percentages of pumps failed by the *ends* of each of the class intervals of Column 1.
3. Three or four possible values, thought likely to span the actual value, are assigned to t_0 (the *guaranteed life*). The resulting values of $t — t_0$ are tabulated in columns 4, 5 and 6. **NB** In each case t is the time of the *end* of the interval, e.g. in Column 4, Row 3

$$t — t_0 = 1300 — 800 = 500 \text{ hours}$$

4. On the Weibull probability paper, the Column 3 figures are plotted first against those in Column 4, then against those in Column 5 and Column 6, respectively.

   The result is shown in Figure 5.13. The value of t_0 which results in the straightest plot, in this case 900 hours, is the one which gives a Weibull cdf which best represents the data.

5. The characteristic life, η, is the value of $t — t_0$ at which the line fitted to the straightest plot reaches the 63 per cent failed level, in this case 600 hours. (**NB** $t — t_0 = 600$ h corresponds to a total actual running time of $t = 1500$ h, remembering that $t_0 = 900$ h.)

6. As shown, a perpendicular is dropped from the fixed 'Estimation point' (printed just above the top left-hand corner of the diagram) to the straightline fit. The point at which this perpendicular intersects the special $\beta—$ scale at the top of the

Table 5.5. Pump failure data

1 *Time to failure* *(hours)*	2 *Number* *of pumps*	3 *Cumulative* *percentage* *failed*	4 $t - t_0$ $t_0 = 800$ h	5 $t - t_0$ $t_0 = 900$ h	6 $t - t_0$ $t_0 = 1000$ h
1000–1100	2	2	300	200	100
1100–1200	6	8	400	300	200
1200–1300	16	24	500	400	300
1300–1400	14	38	600	500	400
1400–1500	26	64	700	600	500
1500–1600	22	86	800	700	600
1600–1700	7	93	900	800	700
1700–1800	6	99	1000	900	800
1800–1900	1	100	1100	1000	900

graph gives the value of β for the best-fit cdf (in this case approximately 3.5, clearly pointing to a wear-out mode of failure).

Small data samples, possibly incomplete, or large multi-censored samples

In practice, it is most often the case that only a handful of times-to-failure have been recorded. Indeed, the components under examination might be large and expensive and only a few might yet have been made. In addition, some of them might still be running, not having reached the failure point (i.e. be *suspended*) or may have been withdrawn from the trial (i.e. *censored*) because, in their case, the test conditions were accidentally altered. In this situation the results of any analysis will necessarily be subject to greater statistical uncertainty. A Weibull analysis might still be needed, however, on the grounds that an approximate result at the end of a fortnight may be of more value than a precise one obtained by waiting for another three months.

In the above case, a technique in which so-called median, or 50 per cent rank, estimations of $F(t)$ values are plotted on Weibull probability paper can then produce meaningful estimates of t_0, η and β. It has the additional advantage that, using published tables of 5 per cent rank and 95 per cent rank estimations, confidence limits can be assigned to the estimated Weibull parameters.

Another situation that can occur is that a relatively large number of failures might have been recorded, but along with data on a large number

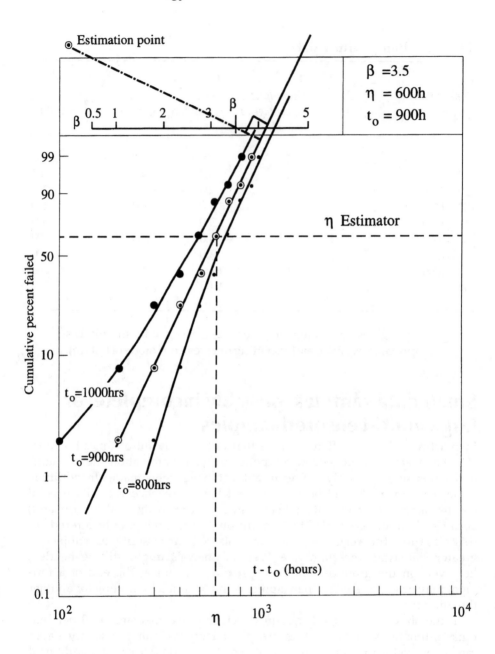

Figure 5.13. Weibull plot of pump failure data, Table 5.5

of suspended or censored items. The censored ones might well, in fact, *have* failed, but by mechanisms different from the failure mode under study. For this situation, $F(t)$ values, for plotting on Weibull paper in the usual way, can be derived from estimates of the so-called *cumulative hazard*. This is a powerful technique; it does not, however, readily lend itself to confidence limit estimation.

Both the median rank and the cumulative hazard techniques can be used to derive separate plots for each mechanism of failure — by taking all failures other than via the mechanism of interest as being censored. This must be done if, as is often the case, several mechanisms, each exhibiting a different β-value, are present. A Weibull plot which lumped all these together would always tend to show, quite misleadingly, a β-value close to unity, leading to the quite erroneous assumption that wear-out, say, was not present.

Although the calculations and graph plotting for these last two techniques are quite straightforward, the statistical reasoning underlying them is sophisticated. Clear, step-by-step explanations, fully illustrated via practical examples and case studies, can be found in the already-mentioned IMechE Guidebook[2], the IMechE's instructional videotape[4] and also in Andrews and Moss[5], a textbook which is strongly recommended to those readers who seek an advanced, but very practically oriented, treatment of reliability analysis.

References

1. Greene, A. E. and Bourne, J., *Reliability Technology*, Wiley, London, 1972.
2. Davidson, J. and Hunsley, C., *The Reliability of Mechanical Systems*, 2nd edn, Mechanical Engineering Publications, IMechE, London, 1994.
3. Murdoch, J. and Barnes, J., *Statistical Tables for Science, Engineering and Management*, Macmillan, London, 1970.
4. IMechE Mechanical Reliability Committee, Video: Learning from failures: Weibull Analysis, technique and applications, obtainable via M. J. Harris, Division of Mechanical Engineering, The School of Engineering, University of Manchester, Oxford Road, Manchester M13 9PL, UK.
5. Andrews, J. D. and Moss, T. R., *Reliability and Risk Assessment*, Longman Scientific, London, 1993.

6
The reliability of plant systems*

Introduction

If we had recorded enough statistical data of the kind that has been described in the previous chapter, and analysed it the ways described — for every one of the components of a complex industrial plant (a chemical process line, say) — then, in principle, we should be able to estimate the expected reliability (the anticipated mean time to failure, for example) of that plant as a whole. It should be possible to assess (a) the reliability of each constituent unit by appropriately combining the reliabilities of its components, and then (b) the reliability of the plant by likewise combining the assessed unit reliabilities (in essence, the 'system reduction' procedure outlined later in this chapter). If, in addition, we had amassed enough dependable data on average times for repairing the likely failures and, of course, data on expected average waiting times (for diagnosis and required resources), we would also be in a position to estimate long-term averages for plant availability, percentage lost production, etc., parameters which are undoubtedly the most basic measures of maintenance performance (see the following chapter, on maintenance objectives, and in particular Figure 7.5 *et seq.*). Unfortunately, both the length and the complexity of such calculations increase exponentially with the number of component items and the complexity of their inter-dependence. All is not lost, however.

Useful insights into probable system performance may often be gained from fairly straightforward calculations which combine the outputs of component reliability observations and analyses, of the kinds that have just been described, with simplified, but sufficiently realistic, modelling of the system. By quantifying the probable influence of the reliability performance of each separate component, or sub-assembly, on that of the plant as a whole such modelling enhances the maintenance manager's feel for the relative importance, or *criticality* of each component, and also for the significance of features of the system design which may render his planning task easier (e.g. unit redundancies) or more difficult (e.g. flow process 'bottlenecks'). Thinking of this kind has already been touched on in Chapter 4, it will be absolutely central to the analysis and development of maintenance strategies which constitutes the remainder of this book (in particular, Chapters 9–12).

---

* Chapter contributed by M. J. Harris, Honorary Fellow, University of Manchester School of Engineering.

As explained, system reliability analysis can be very complex and a great deal of quite advanced effort, in mathematics and in computer methods, has been devoted to it. Much that is most useful and informative can, however, be derived by some relatively straightforward arithmetic, which will now be reviewed. For a more comprehensive introduction, but avoiding complex methods and giving plenty of worked practical examples and tables of useful formulae, the reader is recommended to consult the already-mentioned IMechE Guidebook[1].

Reliability block diagrams

When designing complex engineering plant one of the first tasks is to draw up a block diagram in which each block represents one of the constituent sub-systems or units. This could be a *schematic block diagram*, showing the physical connections, or a *functional block diagram*, showing flows of power, material, etc., with the inputs and outputs specified for each block (for analysis of maintenance strategy the most useful form of the latter is probably a *process flow diagram*, such as that shown in Figure 4.3). At a later stage a related *piping and instrumentation (P&I) diagram* might be constructed.

In a similar manner, the assessment of the overall system reliability may be facilitated by constructing and analysing a *reliability block diagram (RBD)*. In this, the connections symbolize the ways in which the system will function as required and do not necessarily indicate the actual physical connections. Also, for ease of analysis each item is usually modelled as either fully working or totally failed (conservatively, partial failures are assumed to be total ones). An RBD is usually constructed on the basis of the functional, schematic, and/or P&I diagrams.

To illustrate this, consider a system comprising a motor fed by two fuel pumps, both of which are normally working. If the motor is working, full power output will be achieved even if one of the pumps has failed. The RBD would then be as in Figure 6.1, which indicates that power may be achieved via either or both, of the pumps. This would still be the appropriate RBD whether the pumps were located in series, along a single fuel line, or in parallel, provided that full power would still be achieved even with one pump failed and that the failure would not block the fuel line. The form of the RBD is determined by the reliability logic, not simply by the physical layout.

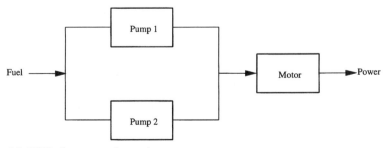

Figure 6.1. RBD of pump–motor system

Series reliability

Consider the simplest system, of just two units. If successful system operation requires both units to be working (i.e. if either one fails the system fails) then for a reliability assessment they are considered to be in series and are represented as in the RBD at the top of Figure 6.2.

If the failure probabilities of the units can be assumed to be *independent* (i.e. the failure behaviour of one is not influenced by that of the other) then the expected *system* reliability at any given time t is given by the product of the two estimated *unit* reliabilities at that time (think of two horses in two separate races: the odds you would be offered against *both* horses winning — a bookmaker's so-called *double* — would be the product of the two separate odds), i.e.

$$R_s(t) = R_1(t) \times R_2(t)$$

Example: If the estimated unit reliabilities at $t = 10\,000$ operating-hours are 0.90 and 0.95, respectively, then the system reliability at that time will be given by:

$$R_s(t) = 0.90 \times 0.95 = 0.855 \text{ or } 85.5\%$$

It can be simply shown that if both types of unit exhibit negative exponential pdfs (i.e. they are in their 'useful-life', random failure phase, see Chapter 5) then the system as a whole will also do so, the system failure rate being simply the sum of the unit failure rates, i.e.

$$f_s(t) = (\lambda_1 + \lambda_2) \exp\{-(\lambda_1 + \lambda_2)\,t\,\}$$

and hence the system reliability will be:

$$R_s(t) = \exp\{-(\lambda_1 + \lambda_2)\,t\,\}$$

and the system mean-time-to-failure

$$(MTTF)_s = 1/(\lambda_1 + \lambda_2)$$

where the λ are the respective unit failure rates.

Example: If

$(MTTF)_1$	$= 100$ h, i.e. $\lambda_1 = 1/100 = 0.01/\text{h}$
$(MTTF)_2$	$= 200$ h, i.e. $\lambda_2 = 1/200 = 0.005/\text{h}$

then, for the system,

$f_s(t)$	$= (0.01 + 0.005) \exp\{-(0.01 + 0.005)t\}$
	$= 0.015 \exp(-0.015 t),$
$R_s(t)$	$= \exp(-0.015 t),$
e.g. at 100 h	
R_s	$= \exp(-0.015 \times 100) = 0.22 \text{ or } 22\%$
$(MTTF)_s$	$= 1/0.015 = 67$ h

All that is required in the case of series reliability systems with more than two units is simple extension of these calculations.

The same general logic applies to calculations of average availability for series systems.

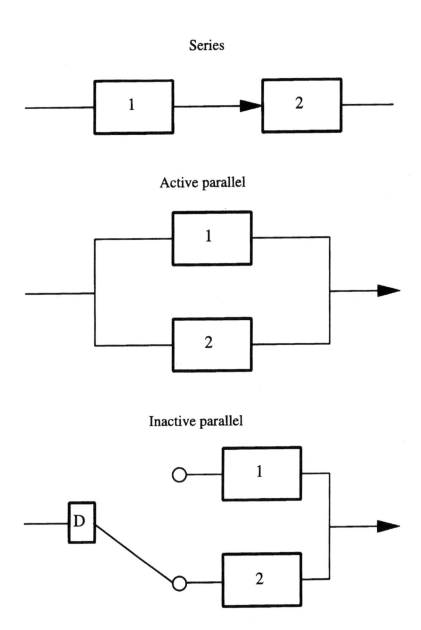

Figure 6.2. Series and parallel connection

Example: A simple process flow system, of four units, is required to run continuously. It has no redundancy (i.e. the units are in series reliability) and the units are subject to randomly occurring failure-and-repair outages which result in average unit availabilities of 98, 98, 96 and 95 per cent, respectively. The system is available only when all units are working. The expected average availability over a long period will be:

$$A_S \quad = 0.98 \times 0.98 \times 96 \times 0.95 = 0.88 \text{ or } 88\%$$

and hence the average unavailability will be:

$$U_S \quad = 1 - A_S = 1 - 0.88 = 0.12 \text{ or } 12\%$$

In this example, because all the unit unavailabilities are very small, e.g.

$$U_1 = 1 - 0.98 = 0.02$$

we could have used the approximation

$$U_S \cong U_1 + U_2 + U_3 + U_4$$
$$= 0.02 + 0.02 + 0.04 + 0.05 = 0.13 \text{ or } 13\%$$

These examples illustrate *Lusser's Rule* — named after the Second World War German rocket engineer who seems to have been the first to formally express it:

The reliability of a series system will always be less than that of its least reliable component.

Active-parallel reliability

Consider, as before, a system of just two units, *both operating*, but in this case joined in such a way that the system only fails if *both* units fail. Such an arrangement is called *active-parallel* and is represented by the second RBD of Figure 6.2. For this, it turns out that the logic is clearer if expressed in terms of failure probabilities $F(t)$, or unavailabilities U (rather than $R(t)$ or A as with series systems). Thus, the system fails only if both units fail so the system failure probability is simply the product of the two unit failure probabilities (*given the same assumptions about statistical independence and so on*), i.e.

$$F_S(t) = F_1(t) \times F_2(t).$$

So

$$R_S(t) = 1 - F_S(t) = 1 - F_1(t) F_2(t) = 1 - \{1 - R_1(t)\} \{1 - R_2(t)\}$$

Example: Two units, active parallel reliability, the unit reliabilities being as in the earlier series example, i.e. at $t = 10\ 000$ operating hours

$$R_1 = 0.90 \text{ and } R_2 = 0.95.$$

The system reliability at 10 000 hours will therefore be:

$$R_S \quad = 1 - (1 - 0.90)(1 - 0.95)$$
$$= 1 - 0.10 \times 0.05 = 0.995 \text{ or } 99.5\%$$

As with series systems, for more than two units the logic is simply extended, i.e.

$$R_S(t) \quad = 1 - \{1 - R_1(t)\}\{1 - R_2(t)\}\{1 - R_3(t)\} \dots$$

If, for the two unit case, both units operate in their 'useful life' negative exponential pdf phase it can be shown that the system pdf will *not* be a simple negative exponential. In fact, it is given by the rather long-winded expression:

$$f_S(t) = \lambda_1 \exp(-\lambda_1 t) + \lambda_2 \exp(-\lambda_2 t) - (\lambda_1 + \lambda_2)\exp\{-(\lambda_1 + \lambda_2)t\}$$

and in larger active-parallel configurations it will be a yet more complex function.

Expressions for MTTF are likewise complicated, e.g. for the two-unit system:

$$(MTTF)_S \quad = (1/\lambda_1) + (1/\lambda_2) - \{1/(\lambda_1 + \lambda_2)\}$$

Example: System as in previous example, i.e. $\lambda_1 = 0.01/\text{h}$ and $\lambda_2 = 0.005/\text{h}$.

$$(MTTF)_S \quad = (1/0.01) + (1/0.005) - \{1/(0.01 + 0.005)\}$$
$$= 100 + 200 - 67 = 233 \text{ h}$$

As before, the same parallel-system logic applies to *availability* calculations.

Example: An active-parallel system comprises three chemical reactors. The system is down only if all three are down. The reactors are subject to randomly occurring failure-and-repair outages which result in average availabilities of 60, 70 and 80 per cent respectively, i.e. average unavailabilities of 40, 30 and 20 per cent — or 0.40, 0.30 and 0.20.

For the system:

unavailability, $U_S \quad = 0.40 \times 0.30 \times 0.20$
$$= 0.024 \text{ or } 2.4\%$$

availability, $A_S \quad = 1 - 0.024 = 0.976 \text{ or } 97.6\%$

Each of these examples illustrates, of course, the converse of Lusser's Rule:

The reliability of a parallel (i.e. redundant) system will always be greater than that of the most reliable of its components.

Active parallel reliability with partial redundancy

In some cases the output from an active parallel system of *several* identical units may be deemed acceptable if at least *some* of the units are working. Such systems are termed *partially redundant*.

Example: An aircraft has four identical, but completely independent, engines. It will continue to fly safely if at least any two engines are working. For each engine the observed mean failure rate (given that it has started successfully) is $\lambda = 0.01/h$, i.e. MTTF = 100 h (not the sort of engine we would entrust life and limb to!). Given that the aircraft has taken off successfully, what is the probability that it will *not* fly safely for 10 hours, assuming that all failures other than engine failure are negligible?

For each engine:

$$R(10\ h) \quad = \exp(-\lambda t)$$

$$= \exp(-0.01 \times 10) = 0.905$$

$$F(10\ h) \quad = 1 - 0.905 = 0.095$$

For the system of four engines:

$$F_s(10\ h) \quad = 0.095^4 + 4 \times 0.905 \times 0.095^3$$

probability of all four engines failing	probability of any three engines failing (there are 4 different ways this can happen)

$$= 8.15 \times 10^{-5} \quad + \quad 3.10 \times 10^{-3}$$
$$= 3.2 \times 10^{-3} \quad \text{or} \quad 0.32\%$$

Inactive parallel, or standby reliability

In much industrial plant, active-parallel arrangements may not be desirable, or even feasible. If, for example, a vessel were to be fed by two positive displacement pumps in active-parallel it might be over-pressurized, so a better arrangement would be to have one pump in-line and the other on standby, as indicated in the lowest diagram on Figure 6.2. Analysis of standby systems, even something as simple as this, can be quite complex. There are several different ways in which this arrangement could fail:

- failure of units 1 and 2;
- failure of D to divert when required;
- D wrongly diverting to a failed unit;
- failure of D to transmit flow.

In addition, there are various repair policies which could be adopted for the failed, off-line, unit so the analysis could also account for:

- random variation in repair time;
- spares waiting time;
- availability of maintenance resources (men, tools and spares);
- repair priority rules, etc.

Clearly, one possibility is that on-line units could fail while the parallel units might still be under or awaiting repair. Analysis can therefore be very complex and is usually accomplished via *Markov discrete state chain analysis* or via *simulation*. Although far from new, these are techniques which are made much more powerful and accessible by computerization and have therefore been applied more widely in recent years; user-friendly specialist packages can now be bought, e.g. the RAMP simulation software sold by TA Consultancy Services, Farnham (availability details for this and other reliability software packages are obtainable from the Safety and Reliability Society[2]).

So far, it has been assumed that the unit or system can be in one of only two states, working or failed. It will often be the case, of course, that some of the units, and hence the system, can operate in a partially reduced state. If the likelihood of this is significant, it has to be taken into account in the reliability calculation, lengthening it still further.

Combining analyses of the kind described into an assessment of the reliability of a system of *many* units, joined in a complex arrangement of process, signal or energy paths, is a lengthy job. However, there are some ways of systematizing the task, as we shall now see.

Reliability analysis of complex or large systems

Truth table

This is actually the least analytical approach. It is readily explained by working through a simple example, which does not really call for this procedure but which will illustrate it effectively.

Example: System as in Figure 6.3. Let X, Y, Z be the respective reliabilities, $R(t)$, of the units at some time t of interest (they could also be the unit average availabilities, if system availability had been the parameter of interest). So

$(1-X), (1-Y)$ and $(1-Z)$ will be the respective failure probabilities $F(t)$.

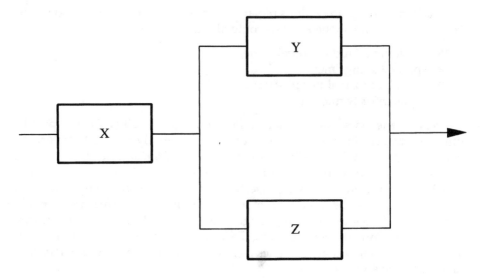

Figure 6.3. Simple system configuration

Denoting *working* status by 1 and *failed* status by 0, we can now construct Table 6.1.

Only the first three states of Table 6.1 are operational ones, so the system reliability is given by the sum of the first three probabilities, i.e.

$$R_s(t) = XYZ + XY(1-Z) + X(1-Y)Z$$
$$= XY + XZ - XYZ$$
$$= R_X(t)\,R_Y(t) + R_X(t)\,R_Z(t) - R_X(t)\,R_Y(t)\,R_Z(t)$$

Separate calculations (not given here) have found that at $t = 20\,h$,

$$R_X = 0.80 \text{ and both } R_Y \text{ and } R_Z = 0.70$$

Table 6.1. Truth table for system of Figure 6.3

Unit status			System status	Probability
X	Y	Z		
1	1	1	1	XYZ
1	1	0	1	XY(1−Z)
1	0	1	1	X(1−Y)Z
1	0	0	0	X(1−Y)(1−Z)
0	1	1	0	(1−X)YZ
0	1	0	0	etc
0	0	1	0	
0	0	0	0	

The expected system reliability at $t = 20$ h will therefore be:

$$R_s(20) = 0.80 \times 0.70 + 0.80 \times 0.70 - 0.80 \times 0.70 \times 0.70$$
$$= 0.73 \text{ or } 73\%$$

The truth table method can be extended to take into account reduced-output states. Analysis of anything other than the smallest systems, however, has to be computerized because even with the two-state model (i.e. each unit can only be either fully operational or fully failed) each additional unit doubles the number of possible system states and hence the computing time. If the units are of high reliability this problem can be alleviated by assuming that the possibility of more than, say, three or four simultaneous failures is effectively zero. One virtue of the method is that it can be used to calculate the reliabilities of systems which have complexities which render ordinary methods of calculation inadequate, a 'cross-over' arrangement, for example (i.e. a path which can divert the flow out of one process line at an intermediate stage, and into a parallel line at the equivalent stage). An illustration of this is given in the IMechE Guidebook[1].

System reduction

The analysis of a large RBD may be carried out in stages, at each of which small groups of units are analysed by the methods so far described, allowing the RBD to be redrawn in a reduced form.

Example: RBD as in Figure 6.4(a), where the number in the bottom right-hand corner of each block is the average availability for that unit.

Stage 1.   Take each series-connected group in Figure 6.4(a), calculate its availability and assign that value to an equivalent single unit, which replaces the original group, i.e.

for groups (2,3), (4,5) and (6,7)
availability $= 0.90 \times 0.80$
                                              $= 0.72$

for groups (8,9,10) and (11,12,13)
availability $= 0.95 \times 0.90 \times 0.80$
                                              $= 0.684.$

Using these results, the RBD can be redrawn as in Figure 6.4(b).

Stage 2.   Take each parallel-connected group in Figure 6.4(b), calculate its availability and assign that value to an equivalent single unit, as in Stage 1.

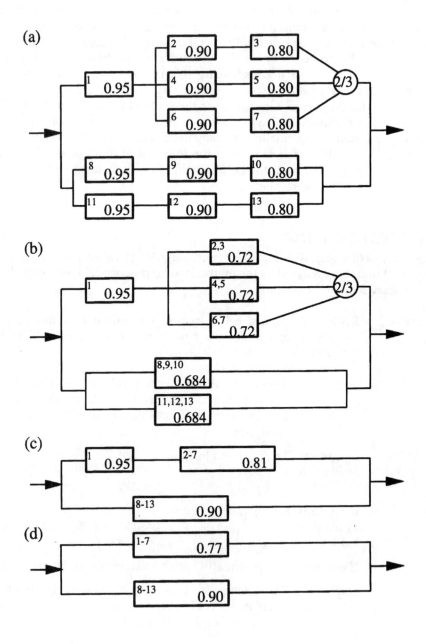

Figure 6.4. Successive system reduction

For group (2–7), operational if at least two channels out of the three are working (i.e. partial redundancy, see earlier),

availability = 0.72^3 + $3 \times 0.72^2 \times (1 - 0.72)$
                  3 channels up   3 channels up, 1 down
                                             (3 ways it can occur)

          = 0.81

For group (8–13), availability = $1 - (1 - 0.684)^2 = 0.90$.

The RBD can therefore be redrawn as in Figure 6.4(c).

Stage 3. On Figure 6.4(c) proceed as in Stage 1.
For group (1–7), availability = $0.95 \times 0.81 = 0.77$.
The RBD is therefore reduced to Figure 6.4(d).

Stage 4. The remaining, parallel-connected, group is now readily analysed, using the formula explained earlier in the chapter.

System availability = $1 - (1 - 0.77)(1 - 0.90)$
                    = 0.98 or 98%

For very large systems, with complex reliability interdependencies, sophisticated *fault tree analysis* (FTA), developed in the 1960s for assessing the reliability of aviation safety systems, is now widely utilized (notably in the nuclear power and chemical process sectors). This is a technique for systematically identifying the various ways in which failure of a single item, or simultaneous failure of several, can in various circumstances induce overall system failure. Again, to be really useful this has needed computerization and many user-friendly commercial packages are now available. An example is the Windows-based 'FaultTree' software, which is designed and marketed by:

Isograph,
10 Mount Street,
Manchester

(see also the SaRS directory[2]).

Fault tree analysis, a *quantitative* technique, needs to be preceded by systematic, and largely *qualitative*, screening of the system concerned to identify (as far as possible) all the significant ways in which system performance could be affected by item failures. Computerized documentation packages to facilitate such a *failure mode effect and criticality analysis* (FMECA) are also available. A very thorough treatment of FTA and FMECA can be found in Andrews and Moss[3].

References

1.   Davidson, J. and Hunsley, C., *The Reliability of Mechanical Systems*,

2nd edn, Mechanical Engineering Publications, IMechE, London, 1994.

2.   *Directory of Software Programs used in Safety and Reliability Assessment,* The Safety and Reliability Society, Clayton House, 59 Piccadilly, Manchester M1 2AQ.

3.   Andrews, J. D. and Moss, T. R., *Reliability and Risk Assessment,* Longman Scientific, London, 1993.

7
Maintenance objectives

Introduction

The previous chapters have stressed the importance of the *maintenance objective*, its starting-point role in the setting of a maintenance strategy and hence the need for its clear formulation. We therefore need to establish what it should consist of, how it can be formulated and how it can then be used — and we will address these matters by looking at them in the context of a real industrial example, in this case an alumina refinery. The discussion, and the necessary plant modelling, will also reinforce the material on plant structure presented in Chapter 4.

Alumina refinery: operating characteristics

Figure 7.1 models the process relationships between: (a) the mine that is the source of the refinery's principal raw material input, namely bauxite; (b) the various material transport systems; (c) the refinery itself; and (d) the refinery's power station. A cursory consideration of the process flow is enough to reveal clearly that, as regards alumina production, the refinery is the rate determining element (although it should be noted that the refinery and the power station are integrated from a production point of view and are therefore both production-critical).

Figure 7.2 shows the process flow of the refinery, modelled at a plant-unit (e.g. bauxite mill) indenture level (for simplicity, the power station is excluded from this representation). Units exhibiting low reliability and resultant high maintenance cost are indicated.

The refinery is basically a sequence of successive processes applied to a circulating working fluid — bauxite and caustic (the main constituent of the fluid) being introduced at its head end, impurities and product being extracted at various stages downstream. The refinery is designed as two identical circuits with spare capacity (stand-by units) either in parallel (as in the case of the bauxite mills), in series (as in the case of the digester banks) or in both series and parallel. In some processes (e.g. the evaporator heat exchangers) there is no stand-by capacity. There is also some interstage storage — available through the use of the precipitators — in the main circuit, but its exploitation incurs a production-loss penalty. Although not part of the cycle, a number of sub-systems spur off, e.g. the hydrate conveyor, kilns and washers. In each case these can be considered to be in series with the main circuit.

The refinery operates continuously — it never comes off line at plant level. Thus any off-line maintenance work has to be scheduled by exploiting the

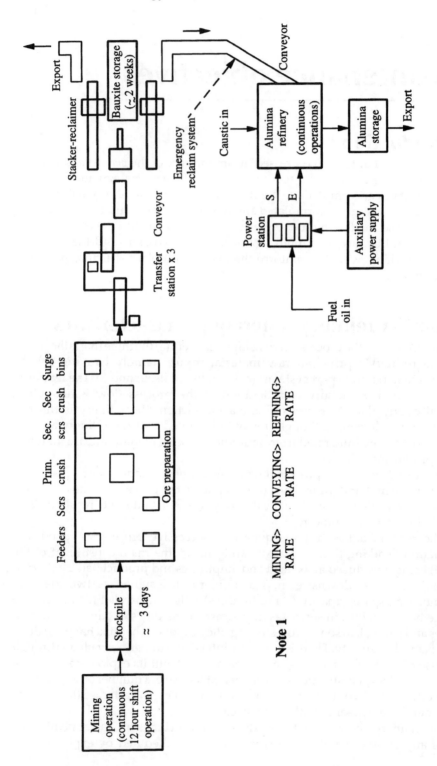

Figure 7.1. Outline process flow, mine and refinery

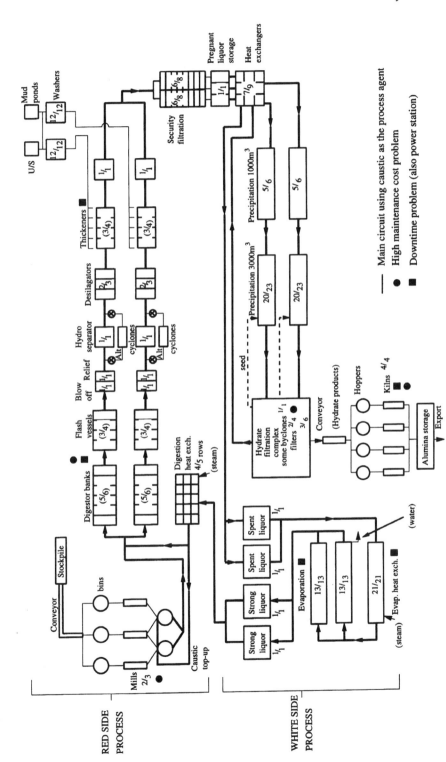

Figure 7.2. Process flow of refinery

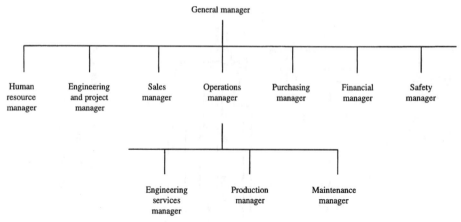

Figure 7.3. Senior management administrative structure

extensive redundancy at unit level. The existing maintenance life plan is based mainly on fixed-time maintenance of the various units. Scheduling is aimed at spreading evenly, throughout the year, the maintenance workload generated by the units. This has resulted in a refinery availability level of approximately 92 per cent.

The senior management *administrative* structure is shown in Figure 7.3 and the maintenance administrative structure in Figure 7.4.

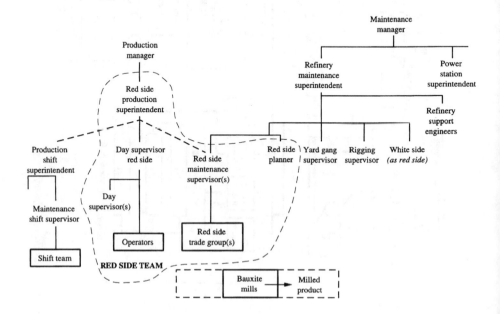

Figure 7.4. Maintenance administrative structure

The maintenance objective

The main factors that should be taken into account in the formulation of a maintenance objective are shown in Figure 7.5. The maintenance resources are used to ensure that plant output (as specified under production policy), safety standards and plant design life are all achieved, and that energy use and raw material consumption are minimized.

In theory, the objective might be considered as being to achieve the optimum balance between the allocation of maintenance resources and the achievement of the plant outputs. In practice, however, the formulation of a maintenance objective is more complex than this. It usually involves the users, owners and safety department specifying what they want from the plant in negotiation with the maintenance department. Only then can the last of these decide how best to maintain the plant (the maintenance strategy) in order to achieve the requirements at minimal maintenance resource cost. This process should provide the basis for maintenance budgeting and cost control.

Maintenance resources and plant output factors

Before considering the above process in more detail it will be instructive to examine the relationships suggested in Figure 7.5, i.e. between maintenance resources and the various factors that determine the nature of plant output.

Maintenance resources (men, spares and tools) — can be considered as inducing the *direct* cost of maintenance. This is relatively easy to measure, using a costing system, and is the maintenance cost the Financial Director is most aware of, what he sees maintenance budgeting as being all about. The maintenance manager can change the level of resources — very quickly in the case of contract arrangements, much more slowly if the resources are in-house.

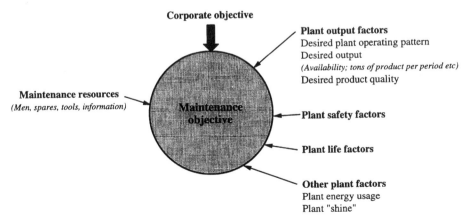

Figure 7.5. Factors influencing maintenance objective setting

Maintenance resources and plant longevity. A proportion of the maintenance resources is necessarily devoted to ensuring that major plant units — and, indeed, the whole plant itself — survive up to or beyond the design operating-life. Failure to ensure this will mean a corresponding loss of capital assets. The maintenance work involved is usually 'protective', e.g. preventing the corrosion of structures, but can also be major part replacement.

Although we all recognize that neglect can cause rapid deterioration, determining the relationship between the level of maintenance and the life of a plant is not easy. The best way of incorporating this into the objective is to establish *standards of plant condition*, which will ensure that the plant will achieve its expected life, and in the light of which the actual plant condition can, periodically, be audited (taking into consideration such factors as obsolescence). Clearly, it is important to identify those parts of the plant that will have a major influence on its longevity. Although it is unlikely that, during the life of a plant, the standards of plant condition will change, the level of maintenance needed to ensure compliance with them will probably increase. One of the problems here is that maintenance that is aimed at prolonging plant life is likely not only to be expensive but also to be needed only very infrequently. Thus, because most costing systems operate on an annual accounting basis, the tendency is often to let such work go until it becomes 'someone else's problem'.

Maintenance resources and desired plant safety (equipment integrity). Here again, there will usually be no clearly appropriate level or frequency of maintenance. The customary procedure is to set safety standards that take account of the estimated probability, consequences and costs of failure. For specific types of plant (e.g. pressure vessels) there are numerous maintenance requirements for ensuring safety — expressed in standards, codes of practice, or legislation. For other items, safety standards will have to be set within the company, although again this will not be easy. Once again the sensible approach is to set safety standards by a process of engineering judgement based on experience and, wherever possible, analyses of plant failures. The extent to which such standards have been complied with should then be periodically audited.

Such a procedure should start with the most senior management. It is their responsibility to understand and comply with the ideas outlined in Figure 7.5, i.e. if there is pressure to cut back on maintenance resources at the same time as efforts are being made to increase output then plant safety standards must not be neglected. This is especially true in the case of ageing and hazardous plants. Many major disasters have stemmed from companies' neglect of the relationship between maintenance resources and safety standards.

Maintenance resources and product output. Product output can be expressed in various ways, see Table 7.1, the most useful of which is usually the output index because it combines several of the other parameters. In the case of the alumina refinery, for example, it is measured in tonnes of alumina — of a particular product mix, and of a defined quality — per period, all these factors determining the company's level of profit.

Table 7.1. Measures of plant output

(i)   Downtime due to maintenance (in hours, per production period, with causes)

(ii)   Downtime index $= \dfrac{\text{downtime per period}}{\text{total planned production time per period}}$

(iii)   Availability index $= \dfrac{\text{uptime per period}}{\text{total planned production time per period}}$

(iv)   Output index $= \dfrac{\text{planned output per period} - \text{lost output per period}}{\text{planned output per period}}$

Planned output per period $=$ planned hours per period x max rate per hour

Lost output per period $=$ (lost production hours per period $\times$ max rate per hour) $+$ (average lost rate per period $\times$ hours $+$ wastage)

A possible relationship between output per period and the level of maintenance resources used is suggested in Figure 7.6. Here, it is assumed that the correct plant maintenance policy is being used, in this case one based on fixed-time maintenance, and that the resources are applied with maximum organizational efficiency. It should be noted that the relationship implies that the cost of achieving increased output rises as the output approaches its maximum. In the case of the alumina refinery the management wanted to increase overall plant availability from 92 to 95 per cent. This would only be worthwhile if the income from the additional sales that would accrue from this increase would be greater than the direct maintenance cost involved in achieving it.

Note also the broken line in Figure 7.6; this indicates the result of imposing a sudden cut in maintenance expenditure. Little effect is felt for the first time period (six months, say) but eventually the output will fall away and money will then have to be spent to bring the plant back to optimal level.

If the cost of lost product is known and is as shown in Figure 7.7 then, in theory and assuming a production limited situation, the optimum operating level can be established. Lost-output cost varies, of course, by orders of magnitude between different industries; in some cases it can be

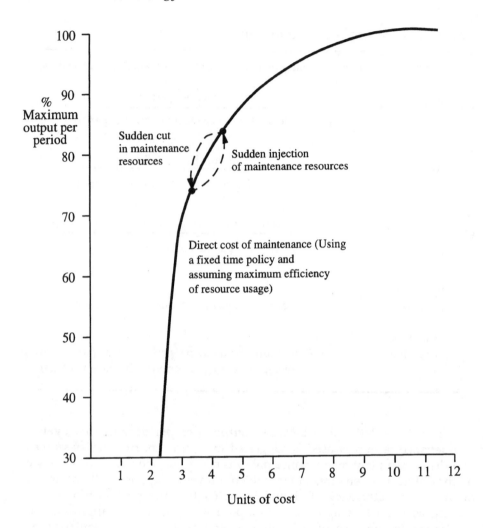

Figure 7.6. Possible relationship between direct cost of maintenance (fixed-time replacement) and output

substantially higher than maintenance resource costs. This pushes the optimum output towards the maximum — see Figure 7.8. In other words, the maintenance objective effectively becomes the maximization of output.

The nature of this relationship can change with the plant's mode of operation, with its age and, in particular, with its maintenance policy. In the case of the alumina refinery, for example, the adoption of a condition-based policy might influence the relationship as indicated in Figure 7.9, i.e. it is not just the level of resources that influences availability but also how they are used.

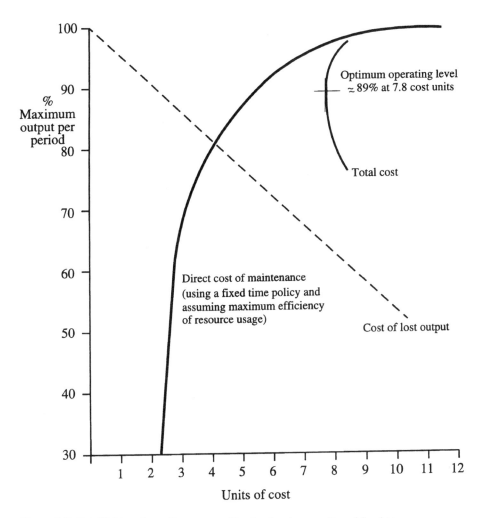

Figure 7.7. Establishing the optimum operating level using a policy of fixed-time maintenance

Probably the main thing to emphasize here is that when considering the overall maintenance life plan for large complex plants the principal focus should initially be not on the various relationships indicated in Figures 7.5 to 7.9 but on directing the maintenance resources towards the most important production units.

The relationships shown in Figures 7.5 to 7.9 have been based on the assumption that the maintenance resources have all been applied with maximum organizational efficiency. Figure 7.10 shows the effect of organizational efficiency on these relationships.

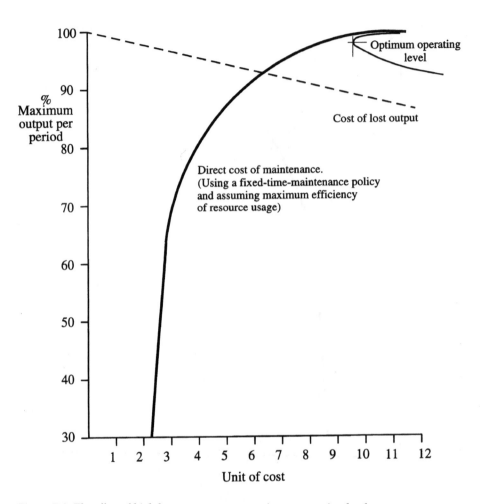

Figure 7.8. The effect of high lost-output cost on optimum operating level

General statement of a plant maintenance objective

Figure 7.5 and the subsequent discussion have identified the main factors concerning maintenance resources and plant outputs that need to be addressed when formulating a maintenance objective. Such considerations suggest the following general statement of the maintenance objective:

> to achieve the *agreed* plant operating pattern, product output and quality, within the accepted plant condition and safety standards, and at *minimum* resource costs

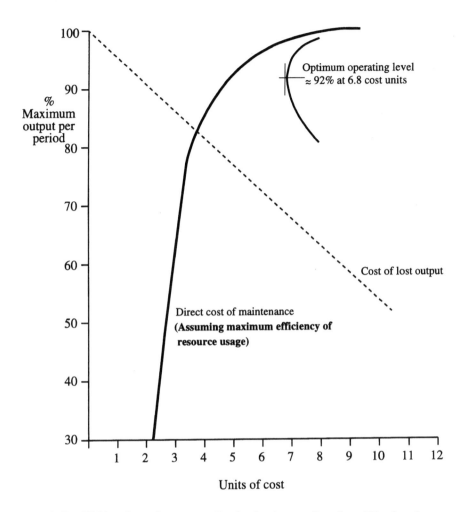

Figure 7.9. Establishing the optimum operating level using a policy of condition-based maintenance

The relative importance of each of the factors included in this statement varies enormously from one technology to another and from one plant to another. With large, production limited, process plants the cost of lost output may be orders of magnitude greater than resource costs; with large buildings the longevity factor may be important, and so on.

Figure 7.11 shows that, while it is necessary to have such an overall maintenance objective, it may be desirable, in practice, to analyse this into sub-objectives concerning each of the output factors and resource areas. The main sub-division is into 'effectiveness' objectives concerning selection of the best life plan (see Figures 7.6 and 7.9) and into objectives concerning the efficient use of resources (see Figure 7.10).

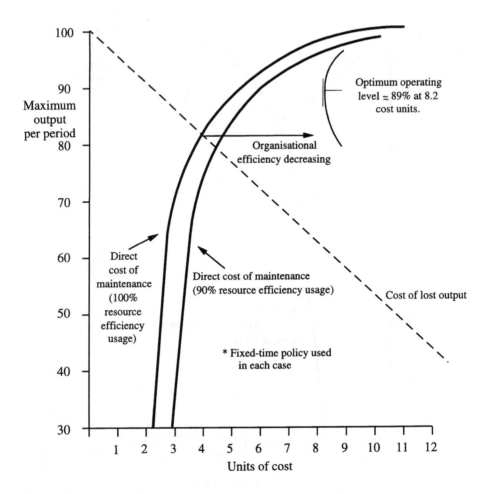

Figure 7.10. Effect of efficiency of resource usage on the location of the optimum operating level

A procedure for formulating maintenance objectives

A procedure is required for establishing maintenance objectives that will be acceptable to the maintenance department, and to all other departments — such as production — whose functions are affected by maintenance. To illustrate how this might be done let us look again at our example, the alumina refinery.

Clearly, the first thing is to get the various 'user departments' to specify what *they* want from the plant. Negotiation to establish these departmental objectives should take place at the refinery's senior management level (see Figure 7.12). If it is to meet the market demand the production department will need to

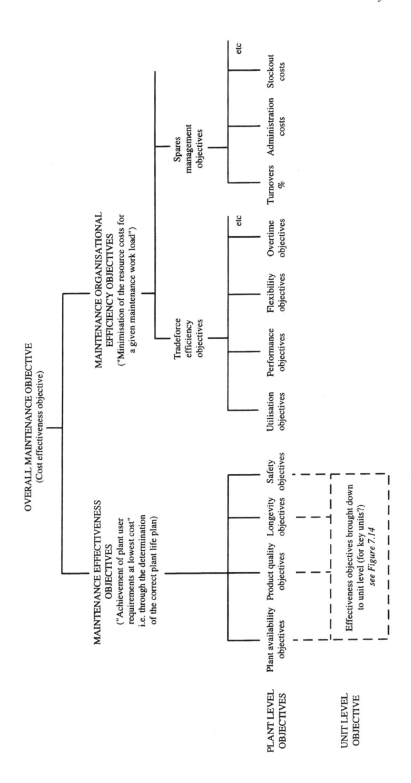

Figure 7.11. Hierarchy of maintenance objectives

establish its operating pattern, its plant availability requirements, and the required product mix and quality. Plant longevity and safety requirements will also have to be identified (see Figure 7.13).

In summary, senior management's maintenance objectives at refinery level are to achieve:

    (a) continuous operation at an average availability of 95%, and
    (b) a plant life of at least thirty years.

There are also quantified targets for personal and environmental safety. All of this to be achieved at lowest cost.

To be meaningful, the above objectives need to be interpreted at plant-unit (e.g. bauxite mill) or production-system (e.g. bauxite milling) level (see Figures 7.4 and 7.14). This can only be accomplished effectively if it is undertaken jointly by production and maintenance, with advice as necessary from those responsible for safety and longevity (where requirements tend to be more unchanging). For example, in the case of the refinery it is the responsibility of the Production and Maintenance Superintendents to establish the production requirement for the milling system (see Figure 7.4) and this might be stated as 'any two out of the three mills should be capable of supplying 100% of the required milled product to the specified milled quality'. The longevity and safety requirements might be for, say, twenty year life expectancy and zero safety incidents.

Taken together, these requirements form the basis of the *user requirement* for the milling system and effectively define the envelope within which it operates. The milling system's maintenance objective — which is compatible with the refinery objective — can then be stated as being 'to achieve the user requirement for the system at lowest cost'. Clear definition of this, and of the user requirement, are a necessary preliminary to establishing the *maintenance life plan* of the system (see following chapter).

In addition to these *plant-oriented* objectives, *organizational* objectives of the type illustrated in Figure 7.11 are desirable. These, if pursued, will improve organizational efficiency and therefore reduce, for a given workload, the direct cost of maintenance. The implication of Figure 7.11 is that having only a *single* organizational objective is impractical. The establishment of objectives for each of the principal maintenance resources — labour and spares — is needed.

The tradeforce objective might be:

    to minimize, for the accomplishment of a given workload, the tradeforce cost per period

This could involve the setting of a target cost, or the setting of sub-objectives expressed in terms of performance targets, utilization levels, flexibility targets, etc. for the tradeforce.

For spare parts the objective could be:

    to minimize the sum of the stockout and holding costs

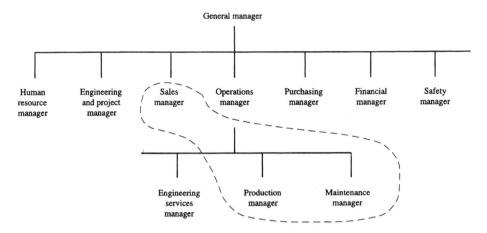

Figure 7.12. Senior management involvement in establishing maintenance objectives

which likewise could be used for setting cost targets or sub-objectives expressed via targets for stock turnover, stockout, etc.

Maintenance objectives in practice

The setting of maintenance objectives is rarely done at all well. More often than not, a written statement of objectives will not exist. Where it does, it is not often based on relationships of the type outlined in Figure 7.6; they are either not understood or ignored. Some of the reasons for this are as follows.

- It is unlikely that data will be available to produce plots such as Figures 7.6 to 7.10 (either at plant or unit level). At best there may be unit level data relating output to the level and type of maintenance resources. In the absence of such information production demand, to meet sales requirements, will dominate the setting of the maintenance objective. This is understand-able because production generates the cash flow and has to be responsive to demand, and to variations in raw material availability, which are often unforeseeable.
- Financial managers often do not appreciate the relationship between maintenance resources and plant factors. They put pressure on the maintenance department to reduce costs without considering the consequences for plant performance and condition.
- Considerable pressure is put on the maintenance department to ensure that mandatory or code-of-practice safety standards are met and that safety-oriented maintenance work is carried out.

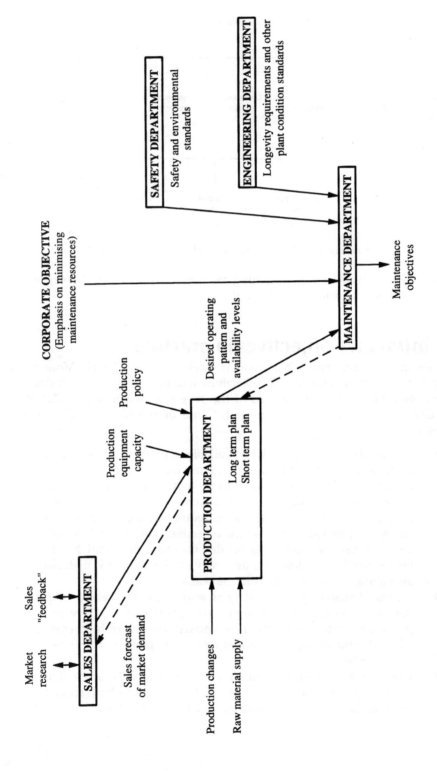

Figure 7.13. A procedure for establishing maintenance objectives

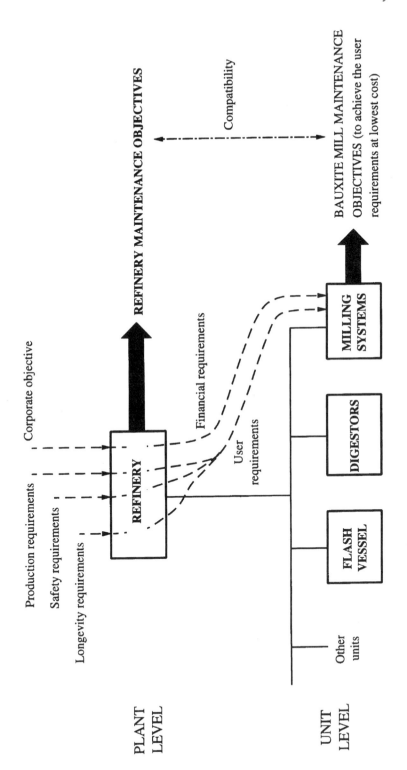

Figure 7.14. Translating objectives down to unit level

Under such pressures the maintenance objective more often than not places emphasis on the short as opposed to the long term. In practice, it is often stated along the lines:

> to achieve the production-demanded output and operating pattern at minimum resource cost, subject to meeting mandatory safety standards

This might be acceptable in the short term but can eventually result in low availability (through neglecting preventive maintenance) and a shorter plant life. Sales, Production and Maintenance must co-operate if a credible maintenance objective is to be established. This will be effective only if the parties concerned appreciate the impact of the relationships modelled in Figures 7.6 to 7.10 and will be enhanced where such models (or even simple trends) are available to aid decision-making.

8
Principles of preventive maintenance

Introduction

Figure 8.1 — a reproduction, for convenience, of Figure 4.10 — is an outline of a typical life plan for a unit of plant. The crucial step in its formulation is establishing the level of preventive work that is needed if the maintenance objective is to be achieved, i.e. 'How much preventive work, and what type, will meet the user requirement at minimum cost?' (see Figure 8.2). Thus, the various possible life plans will be made up of different proportions of preventive and corrective maintenance, ranging from 100% preventive (time-based or condition-based) to 100% corrective.

Figure 8.3 (essentially a development of the model shown in Figure 7.6) shows the relationship, in a production-limited situation, between the level of preventive work and the total maintenance costs. It indicates that there is a level of preventive maintenance that minimizes the sum of the resource costs

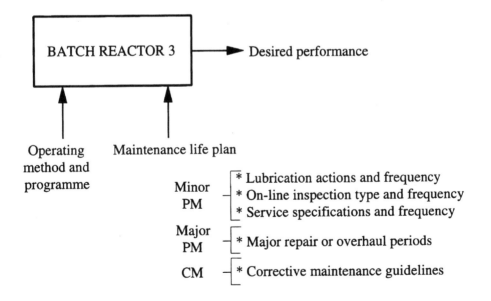

Figure 8.1. A typical unit and its maintenance life plan

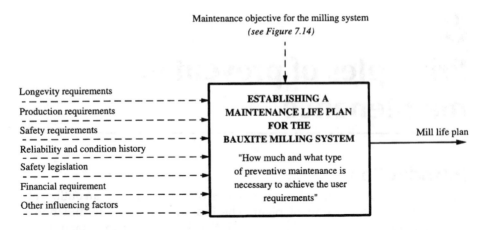

Maintenance objective for the milling system
(see Figure 7.14)

Longevity requirements

Production requirements

Safety requirements

Reliability and condition history

Safety legislation

Financial requirement

Other influencing factors

**ESTABLISHING A
MAINTENANCE LIFE PLAN
FOR THE
BAUXITE MILLING SYSTEM**

"How much and what type
of preventive maintenance is
necessary to achieve the user
requirements"

Mill life plan

Figure 8.2. Factors affecting the maintenance life plan

and the lost output costs. While it may be difficult to precisely locate this minimum it is usually not difficult to determine whether the plant is being under or over-maintained. Models of this type have limited applicability, however. It is not just the frequency (or level) of preventive work that has to be decided but also its nature (time-based, age-based, inspection-based, etc.). In addition, the model assumes that the sales, production, or maintenance position is static, which is rarely the case. For example, the cost of lost output — and hence the optimum level of preventive maintenance — will vary from one time to another depending upon whether the plant is sales limited or production limited. Finally, the model does not take into consideration the complexity of the plant. Some units might be more important than others

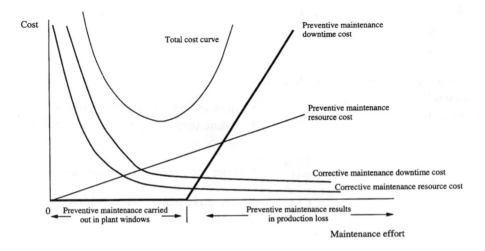

Cost

Total cost curve

Preventive maintenance
downtime cost

Preventive maintenance
resource cost

Corrective maintenance downtime cost

Corrective maintenance resource cost

0   Preventive maintenance carried
out in plant windows

Preventive maintenance results
in production loss

Maintenance effort

Figure 8.3. Relationship between preventive and corrective maintenance costs

(because of the greater impact of their failure on output or safety) and should attract more resources. Indeed, if the model has any applicability at all it is at the lowest level, e.g. deciding on the frequency of replacement of a gearbox. Any procedure for formulating or improving a unit life plan must therefore include the following tasks:

(i)   identifying the maintenance-causing assemblies, sub-assemblies and components that make up a unit;

(ii)   determining the best maintenance procedure for each of the above;

(iii)   assembling the life plan as an amalgam of the selected procedures.

The plant item — a definition

For our purposes we shall define *an item* as being any part of a unit that is likely to require in situ replacement or repair during the life of that unit. Consider, for example, the hierarchical model of the chemical plant shown in Figure 8.4 (a reproduction, for convenience, of Figure 4.5). In this, the complete agitator assembly can be considered as an example of a high level item because at some point in the life of the reaction unit it may be necessary to decide whether to repair or replace it. Conversely, the motor, gearbox, drive belts, coupling, agitator paddle and support frame can also be considered as items. All of these require an in situ repair or replace decision to be taken. As regards the gearbox shaft, however, the decision that is the most likely to have to be made — i.e. whether or not to replace it — will only arise when the box as a whole is in the workshop, so the shaft itself is *not* considered as an item*.

Maintainability diagrams

Replaceability and repair characteristics can be used to construct a *maintainability diagram* such as Figure 8.5, which refers to the chemical plant of Figure 8.4 and which locates the items in a plant hierarchy, i.e. according to their functional dependencies, and also employs a simple code to indicate their repair/replace characteristics. The following categories of item are then identifiable.

- *Simple replaceable* (SRIs, e.g. the drive belt). Likely to be maintained by replacement *in situ* and discard.
- *Complex replaceable* (CRIs, e.g. the gearbox). Likely to be maintained by replacement in situ and repair in workshop.

---

* Solid state electronic equipment is customarily designed with a considerable degree of modularization. Mechanical plant on the other hand, is not often designed for ease of maintenance, and what modularization there is occurs only because of process function and manufacturing method.

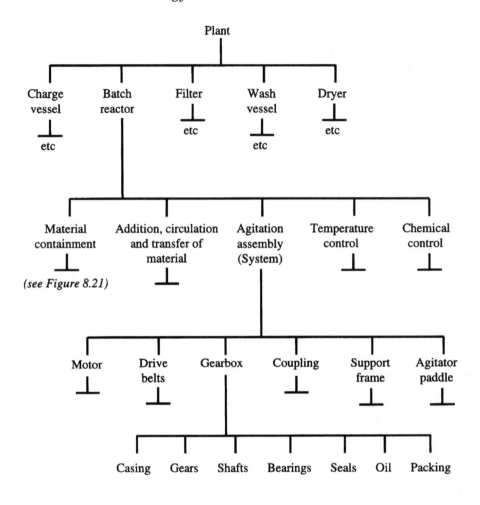

Figure 8.4. Hierarchical division of a batch chemical plant

- *High level* (e.g. the agitator system). Likely to be maintained by repair *in situ* (e.g. agitator shaft welding), but eventually by complete replacement.
- *Special* (e.g. relief valves). Likely to be maintained by periodic proof-testing *in situ* with adjustment, repair or replacement as necessary. These have an intermittent, particular, function only and are usually safety-related. Failure is not observable under normal operation conditions because it has no immediate consequence, but it leaves the plant's integrity seriously vulnerable to some other failure or process deviation, i.e. a consequence which falls in the 'hidden function' category specified in reliability centred maintenance (see Chapter 13).

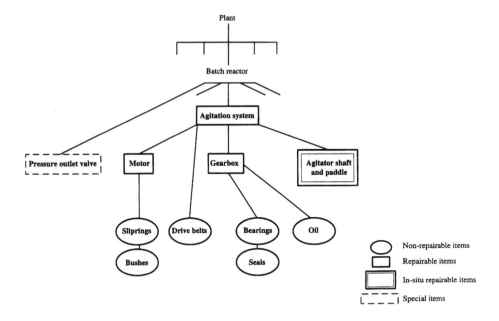

Figure 8.5. Maintainability diagram for a batch reactor

Maintenance procedures and their selection

Preventive maintenance decision-making — the basic task

It can be seen from Figure 8.5 that many maintenance decisions are needed to enable a life plan, of the type outlined in Figure 8.1, to be assembled. Table 8.1 shows the various decision tasks for the gearbox of Figure 8.4. These are the building bricks of the maintenance life plan.

Table 8.1 shows that an item could be *adjusted, calibrated, repaired* or *replaced* and that these actions could be *preventive* (action before failure) or *corrective* (action after failure). Alternatively, the need for such work could be *designed-out*. The main alternative first-level procedures for maintaining an item are listed in more detail in Table 8.2, e.g. periodic repair, periodic visual inspection, and replacement when necessary. Thus, each first-level maintenance procedure consists of:

(i)   the maintenance action,
(ii)   its timing.

To complete this decision scenario additional decisions may be required at second — i.e. workshop — level (e.g. should the removed item be reconditioned or scrapped?). In addition third — stores — level decisions are required, i.e. regarding spares inventory policy.

Table 8.1. The preventive maintenance decision problem

The problem What is the 'best way' to maintain the agitation system gearbox?

THE ALTERNATIVE PROCEDURES	FIRST LEVEL DECISIONS (PLANT LEVEL)		SECOND LEVEL DECISIONS (WORKSHOP LEVEL)	THIRD LEVEL DECISIONS (STORES LEVEL)
	SOME COMBINATION OF:		* Workshop versus contract reconditioning	* Decision to hold *items* or *components* made by repair/replace decision Stores decides on inventory policy, e.g. maximum and minimum stock holdings
	ACTION	**TIMING OF THE ACTION**	* Repair versus recondition versus scrap, items and components	
	* Adjust or calibrate	* Action scheduled *before failure* on usage (hours, miles, etc.) or on calendar time *(fixed -time maintenance)*		
	* Always repair			
	* Always replace	* Action scheduled *before failure* and on condition via inspection *(condition-based maintenance)*		
	* Repair versus replace on condition			
		* Action carried out *after failure,* either unplanned or planned *(operate-to-failure)*		
	OR			
	Re-design, as indicated by failure cause investigation (design-out maintenance)			
Typical influencing information	* Failure characteristics of the item (statistics of failure)		* Cost and quality of internal repair/reconditioning versus contract reconditioning	* Rate of demand for item/ component which in turn is a function of number of such parts in us e in plant
	* Condition monitoring characteristics of the item			
	* Consequences of failure of the item (in terms of safety, production and associated damage)		* Conditioning of removal item/ component and assessment of cost of repair/recondition/new	* Lead time of supply of part and/or is it being internally reconditioned?
	* Resource cost characteristics			
	* Stock holding cost of item compared to stock holding cost of component			* Cost of holding the part versus cost of stock out
	* Downtime cost saving via a replacement versus a repair			

Table 8.2. First-level maintenance procedures

ACTIONS (First level)	TIMING OF THE ACTION
* **Adjust/calibrate/proof test**	* **Based on a policy of fixed-time maintenance (ftm)**. The action is scheduled on usage (hours, miles, output), calendar time, or some combination of these
* **Always repair**	* Based on a policy of **condition-based maintenance (cbm)** or **performance-based maintenance (pbm)**. The timing of the action is based on the condition of the item (or performance) as indicated via one or more of the following inspection methods:
* **Always replace**	

* **Decide on repair versus replace after the failure causing situation has been identified**	* operator functional monitoring
	    * simple inspection (look, listen, feel, smell)
	    * condition checking (against a limit)
	    * trend monitoring
	
	        (a) at fixed intervals or
	        (b) at variable intervals or
	        (c) via continuous inspection or
	        (d) via some combination of these

	* Based on a policy of **operate-to-failure (o tf)**. The action is carried out after failure. The action can involve considerable pre-planning (via spares, quick change) if the item is designated as critical
	* Based on a policy of **opportunity maintenance (om)**. The timing of the action is based on some other item's maintenance timing

Design-out maintenance — as indicated by failure cause investigation after major or recurring failures

The maintenance actions

Primarily these are as follows:

- *Adjustment* (or *calibration*). Carried out with the aim of compensating for some ageing mechanism, bringing an item's function back within prescribed limits. Can be regarded as largely independent of the other actions and can be considered separately.
- *Proof testing*. Checking the operational capability of special items or units. Also independent of the other actions and can be considered separately.
- *Replacement*. The maintenance of a unit by the replacement of its constituent items.
- *Repair*. The maintenance of a unit by the in situ replacement or repair of the constituent components of an item.

It can be seen from Table 8.1 and Table 8.2 that, in practice, the available decision options for an item of plant are:

- (i)   always repair the item in situ (on-line or off-line and before or after failure);
- (ii)   always replace the item (on-line or off-line and before or after failure);
- (iii) leave the replace versus repair decision until item maintenance has been deemed necessary, e.g. the failure-causing event has occurred.

The main factors to be taken into consideration when deciding on which of these is best are:

- the repair and replacement characteristics of an item as indicated in Figure 8.5;
- the extra costs of holding both an item *and* its components (for replacement) rather than just the components (for repair);
- the possible saving in downtime costs if item replacement is speedier than item repair.

In most practical situations the comparison of these main factors, coupled with engineering judgement, allows the best of options (i) to (iii) to be identified. For example (see Figure 8.5), the only option for SRIs is to replace in situ and discard; the most likely option for many CRIs, such as electric motors, is to replace and send to internal or contract workshop for reconditioning. However, with some high-level, high-cost items (such as the agitator) the selection of the best action may not be straightforward; it may be necessary to analyse each of the above first-level actions into its decision scenario (see below) in order to carry out a full cost comparison of the options.

Always repair. Where only in situ repair is feasible the decision scenario is as shown in Figure 8.6. Item repair is only possible if a component is in stock or can be quickly bought in. Stores policy

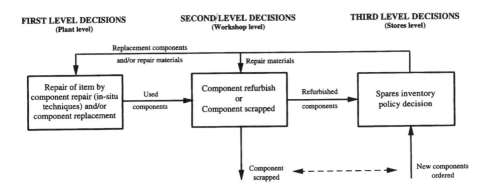

Figure 8.6. Decision scenario if the first-level decision is 'repair item'

might be to hold any component that is likely to be required during the life of the plant. The rationale used for assessing the optimum number of components to hold and the optimum time and quantity for re-ordering is known as the *spares inventory policy*, which will take account of the rate of demand for the component (and therefore of the number of such components in use in the plant). In some situations it may well be economic to refurbish the component.

Always replace. For this to be possible the item would need to be held in stores. If reconditioning were to be carried out internally the components also would have to be held. Where only item replacement is feasible as a first-level decision the scenario is as shown in Figure 8.7. Here it is assumed that the item repair is carried out internally and the workshop decisions involve choosing between repair, recondition or scrap, a decision to scrap having consequences for the stores inventory. Such items are sometimes referred to as *rotables* (see Figure 8.9).

Repair versus replace. The first-line replace/repair decision is sometimes left until the failure has occurred or is imminent. Such a policy would be adopted partly because of the high cost of the replacement work and partly because of the wide range of possibilities for the type of failure that would occur — each type requiring a different method of repair (see Figure 8.8). The decision would be influenced by such information as:

- probable defective part and *in situ* repair methods available;
- time, labour and material cost of item replacement;
- unit unavailability cost;
- running time to next 'window of opportunity'
- probable life of item after repair or temporary repair;
- probable life of item after replacement;
- condition and probable life of unit.

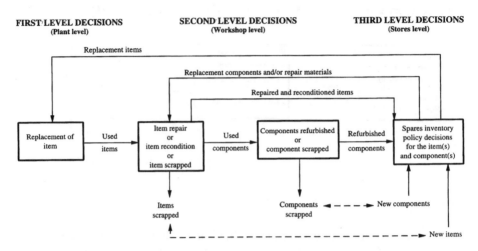

Figure 8.7. Decision scenario if the first-level decision is 'replace item'

Clearly this is *dynamic* decision-making which would be aiming, as far as possible, at cost minimization but would also call for judgement of non-quantifiable factors. It is made much easier if some form of inspection procedure has provided prior warning of failure. If this is not possible and the item is considered critical then decision guidelines (with job procedures) must be established. Choosing, in a dynamic situation, between the repair or replacement of a complex item is the most difficult, and the most commonly occurring, maintenance decision-making problem.

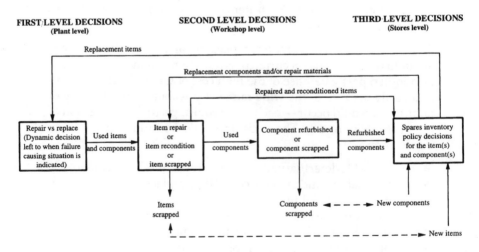

Figure 8.8. Decision scenario for a dynamic first-level 'replace versus repair' choice

Reconditioning — internal versus contract

Although not indicated in Figures 8.6 to 8.8, an associated and secondary question is whether to maintain the repairable items and components internally — by setting up an in-plant workshop — or to use contract repair or exchange (or some combination of these). Although this would usually be decided on economic grounds there could well be other influencing factors, i.e.

- the availability of contractors;
- the complexity of the repair;
- quality assurance needs;
- security of supply.

Most companies have some combination of internal and contract repair. Controlling this can be one of the most difficult maintenance management problems. Figure 8.9 maps a typical system, for a large process plant, for dealing with repairable items.

In situ repair techniques and the repair versus replace decision

In situ repair techniques are to *corrective* maintenance what condition based maintenance (CBM) techniques are to *preventive* maintenance, although the former have received far less attention in the open literature. For a maintenance technique to be considered as an *in situ* one, all phases of the repair process must be undertaken at the item's normal location. In situ techniques have major advantages where the cost of downtime is high. Their adoption can extend running times and improve availability via the reduction of unit outage times, and changes the balance of the trade-off outlined on page 104, many more items being repaired rather than replaced, or at least replaced less often.

Figure 8.10 details an on-line in situ technique for valve replacement. Appendix 2 lists the principal in situ techniques.

The timing of the maintenance action

Fixed-time maintenance (FTM)

These are first-level actions (repair or replace) carried out at regular intervals, or after a fixed cumulative output, fixed number of cycles of operation, etc. They include item replacement, item repair and major strip-down for inspection (the author regards condition-based maintenance as inspection carried out without major strip-down).

FTM will improve reliability only if the failure of the item concerned is clearly a result of some form of wear-out and if the useful life of the item is less than that of the unit in which it belongs. Consider, for example, the case modelled

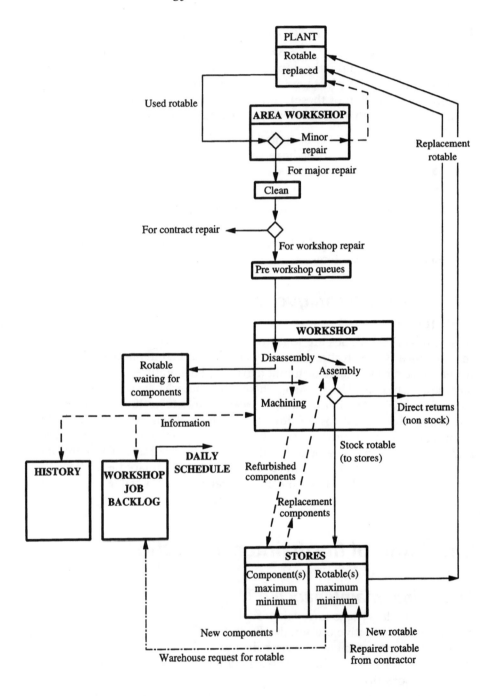

Figure 8.9. System for reconditioning rotables

- The valve to be changed is closed and downstream piping is disconnected at the valve.

- The tool is attached to the valve, *see Figure (a)*

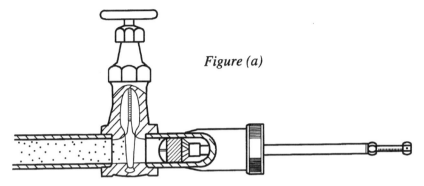

Figure (a)

- The valve is opened.
- The tool is pushed through the open valve, *see Figure (b)* and the seal is expanded until it presses tightly against the pipe wall and seals it.

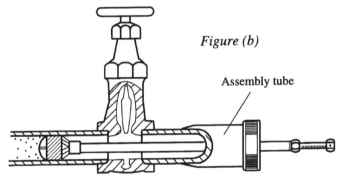

Figure (b)

Assembly tube

- Assembly tube A is removed, the pipe remains sealed, *see Figure (c).*

- The faulty valve can then be removed for repair or replacement.

Figure (c)

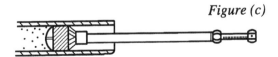

- A new valve is fitted and assembly tube A is reattached. The seal is released and the tool withdrawn. The valve is closed and the tool removed. The downstream piping is replaced.

Figure 8.10. On-line valve replacement — principles of procedure

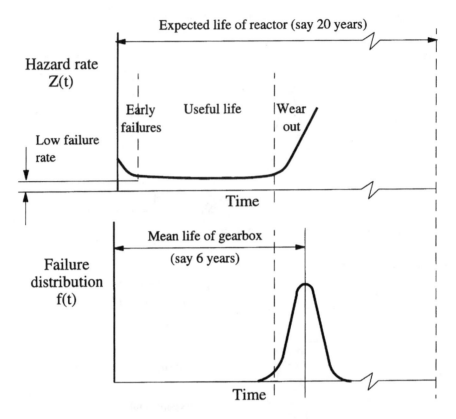

Figure 8.11. Reactor and gearbox failure statistics

in Figure 8.11. The reaction vessel has a design life of approximately twenty years, while the gearbox of its agitator system is likely to wear-out after about six years. FTM (i.e. replacement before six years has elapsed) of the gearbox would be an effective policy for improving the reliability of the agitator system.

The cost-effectiveness of fixed-time maintenance will depend (among other things) on the predictability of the time-to-failure of the item concerned, i.e. on the dispersion, or spread, of the distribution of observed times-to-failure. The smaller the relative dispersion — and hence the larger the Weibull β-factor — the greater the predictability (see Chapter 5). For example, the time-to-failure of the motor in Figure 8.12 is much more uncertain than that of the gearbox and the fixed-time maintenance of the former would be more difficult to justify. In practice, of course, the big difficulty in applying this kind of analysis is that the statistical data is not often available.

Figure 8.13(a) shows the failure distribution for an item whose useful life exceeds that of the system of which it is a part. Thus the probability of its failure during the life of the system is low and the incidence of such failures random. FTM is therefore *not* an effective policy for improving reliability. The same can

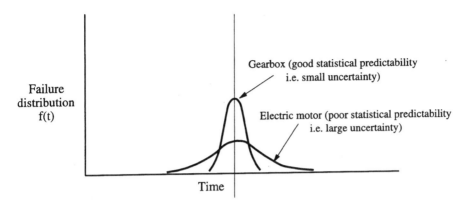

Figure 8.12. Statistical predictability

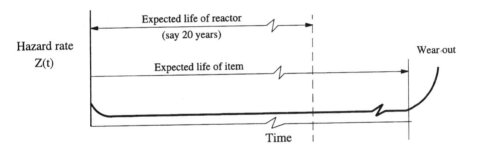

(a)

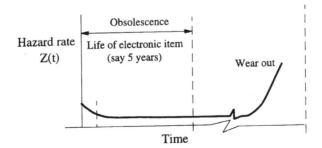

(b)

Figure 8.13 (a) Item in useful life stage during life of reactor. (b) Item replaced due to obsolescence before wear-out

be said for the electronic item of Figure 8.13(b). If, in such cases, failure is unacceptable (or undesirable) then the alternative policies of Table 8.2 must be considered.

In spite of these limitations FTM, in one of the following forms, is often the most appropriate policy:

(i)   *Group (or phased) replacement of large populations of identical (or similar) items (e.g. lamps, overhead cable supports, etc.).* Consider, for example, the maintenance, in a large building, of lamps the failure distribution of which might be as in Figure 8.14. One policy might be to replace them all annually, whether failed or not, and at worst only 10 per cent of them would have failed before this renewal. Such a policy is usually much cheaper — in terms of the combined cost of labour and materials — than the alternative of replacing each lamp as it fails.

(ii)   *Group replacement of dissimilar items (usually the SRIs — see Figure 8.5) in a service period or in a window of opportunity presented by a break in production.* The idea is indicated in Figure 8.15 where those simple items whose failure statistics show that they will last at least 12 months but probably not 24 months are replaced as a group at 12 months. Such a policy might well ensure that an acceptable plant reliability would be achieved at a more affordable cost.

(iii)   *The replacement of safety-critical or production-critical repairable items (usually the CRIs, e.g. gearboxes, see Figure 8.5) where on-line inspection is not possible and the plant runs for long periods without shutdown.* The main difficulty is deciding on the item's replacement period. In most practical situations there is uncertainty caused by the lack of failure data. The best policy is probably to play safe and replace at cautiously short operating periods which can be extended with experience and in the light of information gained via inspection of any items removed for, say, reconditioning (see Figure 8.16).

(iv)   *The shutdown of major process plant for overhaul,* where a large contract tradeforce and other external resources are required. Note that one of the prime uses of condition-based maintenance (see following section) is to provide information to enable the shutdown workload to be predicted and planned in advance of the shutdown.

Condition-based maintenance (CBM)

An attractive idea is that the appropriate time for maintenance ought to be determinable by monitoring condition or performance — provided, of course, that a readily monitorable parameter of deterioration can be found.

Let us assume, for example, that the vibration level of the gearbox of Figure 8.11 can be monitored, and a failure event thus forecasted (see Figure 8.17), with a lead time of about five months. A monthly inspection interval will

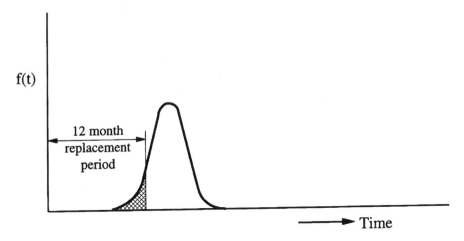

Figure 8.14. Lamp group replacement based on failure distribution

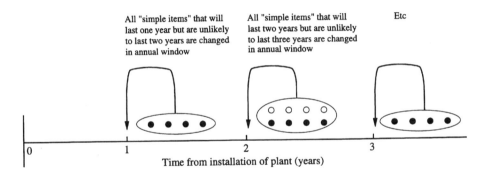

All "simple items" that will last one year but are unlikely to last two years are changed in annual window

All "simple items" that will last two years but are unlikely to last three years are changed in annual window

Etc

Time from installation of plant (years)

Figure 8.15. Group replacement of dissimilar items

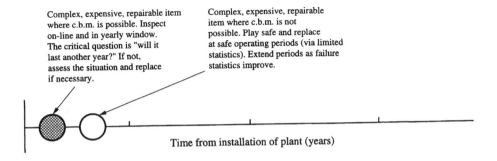

Complex, expensive, repairable item where c.b.m. is possible. Inspect on-line and in yearly window. The critical question is "will it last another year?" If not, assess the situation and replace if necessary.

Complex, expensive, repairable item where c.b.m. is not possible. Play safe and replace at safe operating periods (via limited statistics). Extend periods as failure statistics improve.

Time from installation of plant (years)

Figure 8.16. Policies for complex repairable items

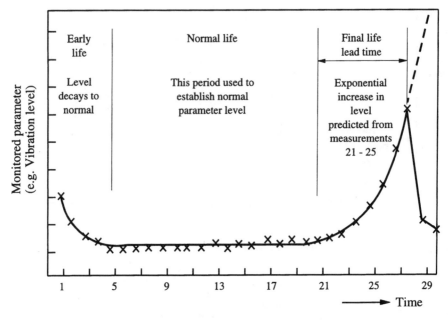

Figure 8.17. Vibration monitoring of gearbox

give adequate notice for planning and scheduling of the maintenance, thus minimizing the effect of lost production. The advantage of this over FTM is that it allows the operation of *individual items* up to nearly their maximum running time.

CBM is also the prime policy where an item exhibits little failure predictability (see Figure 8.12), or fails randomly (see Figure 8.13), or the statistical failure data is not available. It is particularly important for expensive repairable items (see Figure 8.16).

Clearly, the inspection interval adopted will be determined by the lead time to failure. The effectiveness of CBM depends in no small measure on how reliably this can be determined from deterioration curves of the type shown in Figure 8.17.

The monitored parameter can provide information about a single component (e.g. about the wear of a brake pad) or about changes in any number of different components. The more specific the information provided, the better it is for maintenance decision-making.

Condition monitoring can be accomplished in three main ways:

Simple inspection.   Mainly qualitative checks based on looking, listening and feeling (e.g. to detect rope wear). The cost of this should be insignificant compared to the cost of replacement or repair. The period between inspections should be sufficiently short so that minor problems can be detected before they develop.

| Condition checking. | Done routinely by measuring some parameter which is not recorded but is used for comparison with a control limit. Such checking only has value where there is extensive experience of identical systems. |
| Trend monitoring. | Measurement and graphical plotting of a performance or condition parameter in order to detect gradual departure from a norm (see Figure 8.17). This application is most effective where little is known about the deterioration characteristics. When enough knowledge of these has been acquired, condition checking can be substituted for trend monitoring. |

The monitoring of performance and of condition are closely related. In the former case the parameter monitored would be some measure of a unit or item input or output, e.g. power consumed, pressure delivered. Changes in this might then be related to deterioration in some condition of concern. For example, Figure 8.18 shows how changes in performance of a pump can indicate a particular type of wear.

Where CBM is used extensively the high level of data collection renders mandatory the use of computerized systems of the type shown in Figure 8.19. These often have trend analysis capability. The extensive range of monitoring techniques available is indicated in Table 8.3.

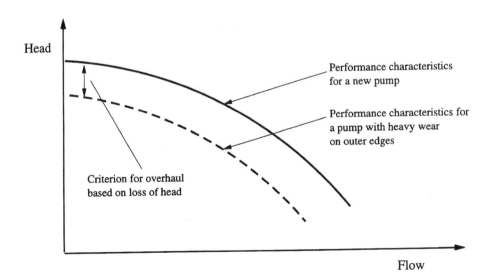

Figure 8.18. Performance monitoring criterion for centrifugal pump

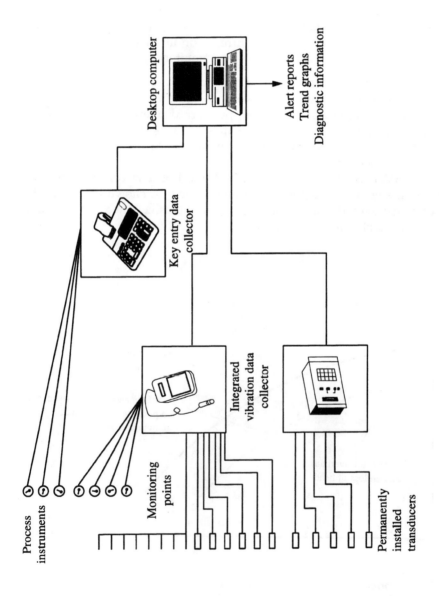

Figure 8.19. Collection of condition monitoring data

Table 8.3. Condition monitoring techniques

Type	Method	On/off line	Manual/ automated	Comments	Skill of operator	Equipment cost (£)
Visual	Human eye	On/Off	Manual	Covers a wide range of highly effective condition checking and surface inspection methods	Low/ Average	0
	Optical probes	Off	Manual	Can be used for internal inspection of machines, good for detecting surface corrosion, wear and severe defects like cracks	Average	50–1000
	Closed circuit television (CCTV)	Off	Manual	Permits detailed inspection of inaccessible/hazardous environment machine parts. Image recording and high resolution analysis is a post-processing possibility	High	1000–0000+
Temperature	Temperature crayons, paints and tapes	On	Manual	Simple and effective aids to visual inspection. Can resolve body temperature to within a few degrees and monitoring can be performed from a distance at a glance	Low	10–50
	Thermometers, thermocouples	On	Manual/ Automated	Range from stick-on thermometric strips to permanently installed thermocouple sensors. Can give visual temperature readout or an electrical input to a hard-wired monitoring system	Average	20–500+
	Infra-red meter	On	Manual	Non-contacting device which measures radiated body heat to estimate the surface temperature of a component. Covers a wide range of temperature but acts only on a small area	High	250–3000

Table 8.3. Condition monitoring techniques *(cont'd)*

Type	Method	On/off line	Manual/ automated	Comments	Skill of operator	Equipment cost (£)
	Infra-red camera	On	Manual	As above but can cover a much wider surface area. Can provide a detailed surface temperature and can be calibrated to give quantitative measurement	High	1000–7500
Lubricant	Magnetic plugs and filters	On/Off	Manual	Analysis of debris picked up by plugs or filter in an oil washed system. Mainly large debris picked up, 100–1000 microns	High	50–1000
	Ferrography	N/A	Manual	Analytical technique used to separate ferrous debris by size to enable microscopic examination. Non-ferrous debris can also be separated but not graded. A wide range of debris size can be analysed from 3–100μm. A contract service is usually available	High	12 000+ (for the machine)
	Spectroscopy	N/A	Manual	Analytical technique used to determine the chemical composition of the oil and debris. Generally for small debris size 0–10 microns. A contract service usually available	High	14 000+ (for the machine)
Vibration	Overall vibration level	On	Manual/ Automated	Represents the vibration of a rotating or reciprocatingmachine as a single number, which can be trended and used as a basis for the detection of common machine faults, but fault diagnosis is not possible and detection capability can be compromised	Average	150–1000 (for a hand held data collector)

Table 8.3. Condition monitoring techniques *(cont'd)*

Type	Method	On/off line	Manual/ Automated	Comments	Skill of operator	Equipment cost (£)
	Frequency (spectrum) analysis	On	Manual/ Automated	Represents the vibration of a rotating or reciprocating machine as a frequency spectrum (or signature) which reveals the discrete frequency component content of the vibration. Provides the basis for fault detection, detailed diagnosis and severity assessment	Expert	10 000+ (for a hand held data collector)
	Shock pulse monitoring (SPM), spike energy and kurtosis	On	Manual/ Automated	All of these techniques use high frequency vibration signals to detect and diagnose a range of faults including rolling element bearing damage, lubrication failure and leak detection	High	400–2500
	Structural monitoring	Off	Manual	A variety of vibration-based techniques exists for the detection and location of structural faults. The majority of such techniques involve imparting a known vibration into the structure and analysing the resulting response	Expert	15 000+
Crack	Dye penetrant	On/Off	Manual	Detects cracks which break the surface of the material	Average	10–150
	Magnetic flux	On/Off	Manual	Detects cracks at/near the surface of ferrous materials	Average	100–500
	Electrical resistance	On/Off	Manual	Detects cracks at/near the surface and can be used to estimate depth of crack	High	200–1000

Table 8.3. Condition monitoring techniques *(cont'd)*

Type	Method	On/off line	Manual/ automated	Comments	Skill of operator	Equipment cost (£)
	Eddy current	On/Off	Manual	Detects cracks near to surface. Also useful for detection of inclusions and hardness changes	Expert	250–5000
	Ultrasonic	On/Off	Manual	Detects cracks anywhere in a component. Suffers from directional sensitivity, meaning that general searches can be lengthy. Often used to back up other NDT techniques	Expert	500+
	Radiography	Off		Detects cracks and inclusions anywhere in a component, although access to both sides of component is necessary. Involves a radiation hazard	Expert	8000+
Corrosion	Weight loss coupons	Off	Manual	Coupons are weighed and weight loss is equated to material thickness loss due to corrosion	Low	25–100
	Incremental bore holes	On	Manual	A series of fine plugged holes of incremental depths which are periodically unplugged and scrutinized for leakage		
	Electrical resistance	On	Manual/ Automated	Electrical element and potentiometer are used to assess resistance change due to material loss. Capable of detecting material thickness reduction of less than 1 nm	Average	100–750
	Polarization resistance	On		A good indicator of corrosion but is unreliable as a means of estimating material loss rate	Average	150–500

Operate to failure (OTF)

No action is taken to detect the onset of, or to prevent, failure. The resulting demand for corrective work occurs with little or no warning. This will only be cost effective if:

(i)   the consequences of item failure in terms of lost production, or of danger or damage, can be regarded as negligible (or, alternatively, if the cost of letting the item fail is less than that of implementing alternative maintenance policies);

(ii) the consequences of item failure are serious but do not take effect for some time and it is possible to carry out the necessary repair within this period. (Obviously, such failures have to be identified when formulating the maintenance life plan.)

Opportunity maintenance (OM)

The timing is determined by that of some other action, e.g. much power station plant is maintained during the statutory boiler inspection (see Chapter 9 for a fuller discussion of this).

Design-out maintenance (DOM)

By contrast with the previous policies, which are aimed at the avoidance or mitigation of failure, design-out aims to eliminate the cause of maintenance altogether. Clearly, this is a design problem but it is often part of the maintenance department's responsibility. In general, DOM only becomes an alternative after some experience of operation, i.e. if, in spite of carrying out the original maintenance procedure, the performance of the item remains unsatisfactory (exhibiting low reliability or high maintenance cost). Therefore, such a policy can only be implemented effectively if an information system exists which facilitates the choice (on cost and safety grounds) between the proposed redesign and the best of the various maintenance procedures (see Chapter 10).

Guidelines for establishing the best timing of a maintenance action

In the earlier part of this chapter we discussed the problem of deciding the best maintenance *action*. We now turn to the related question of *timing*. The recommended logic for doing this is encapsulated in the decision chart shown in Figure 8.20.

The first step for any item under consideration is to decide whether it is a *special* one as defined earlier. If it is, then the usual procedure, as already explained, is to proof test it at fixed time intervals, assessing the appropriate duration of which is a complex business which involves taking into account[1]:

(a) both the fail-to-danger and fail-safe rates of the item itself;

(b) the anticipated incidence of the process deviation (e.g. over-pressure) which the item is protecting against;

(c) the time required to carry out the proof test (during which the item's protective function may be disarmed);

(d) the degree of item redundancy and 'majority voting' arrangements, and hence the possibility of staggered testing;

(e) (last but not least) the safety, environmental and economic consequences of an occurrence of the unprotected process deviation (i.e. of the feared accident).

In addition, each special item should be subject to the same decision logic (Figure 8.20) as normal items, so that a pressure relief valve may not only be tested fairly frequently (every few weeks, say) but also replaced at longer intervals (e.g. every few years).

For a normally operating item the logic of the chart first leads to the identification of the various effective actions (in the light of the failure characteristics listed in Table 8.4) and then to the identification of the best of these (in the light of the cost and safety characteristics listed in Table 8.5).

In the majority of cases the best procedure can be identified via engineering judgement of the main influencing factors and application of the logic of Figure 8.20. For example, the initial ranking given in the figure is as follows:

1. Condition-based maintenance (on-line).
2. Condition-based maintenance (off-line).
3. Fixed-time maintenance.
4. Design-out maintenance.
5. Operate to failure.

Table 8.4. The main failure characteristics (see also Chapter 5)

- Useful life
- Mean life
- Uncertainty of item life (is the statistical predictability good? See Figure 8.12) as measured by range, standard deviation, Weibull β
- Detectability of the onset of failure (is inspection possible?): parameter to be monitored, techniques available, lead time to failure
- Mean replacement time
- Repairability (can the item be reconditioned?)

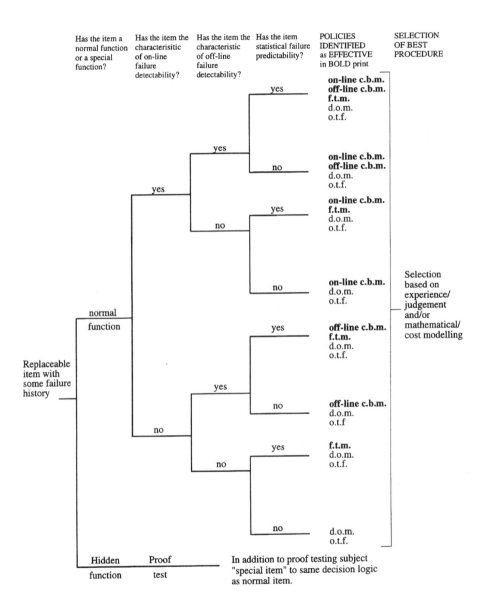

Figure 8.20. Decision logic for identifying the best maintenance procedure

Table 8.5. The main cost and safety characteristics

- Consequence of failure in terms of cost output, poor product quality, gearbox running costs, associated damage, increased maintenance resource cost
  Consequence of failure in terms of actual or potential hazards to personnel and/or general public
  Consequence of failure in terms of environmental damage

Requires an understanding of
- the item's function
- the ways in which the item's functional performance can be lost or reduced

- Statutory safety requirements
- Direct costs of maintenance, including:
    Cost of item (to purchase and to hold a spare)
    Cost of labour to replace and/or repair
    Cost of the best inspection technique (if any)
    Cost of re-design (if possible)

This can be justified on the following grounds:

- If an effective on-line condition-based procedure can be found it is normally the best one to adopt, especially in process plants, where downtime cost is usually very high.
- CBM is usually cheaper and more effective than FTM because of the inherent variability of the item time-to-failure or lack of knowledge about it.
- FTM is normally cheaper than OTF because of the high cost, in the latter case, of the lost production or the consequent damage.

In only a few situations will the maintenance manager need to employ statistical reliability or cost analysis to determine the optimum procedure.

Examples of maintenance procedure selection

1. *The rubber lining of a chemical reaction vessel* (see Figures 8.4 and 8.21)

The main work needed on these vessels is the repair (needing 2½ days work) or replacement (needing ten days work) of the rubber lining, which is subject to permeation by chemicals and which can eventually deteriorate to failure. It has been found that because this process can be accelerated by accidental damage the time-to-failure of the lining can be anything from three to nine years, with a mean value of six years. The deterioration is, however, slow and gives about a year's notice of failure. It is important not to allow the rubber to deteriorate too far. Penetration of the steel vessel is hazardous and also necessitates a long and expensive repair. If the deterioration can be

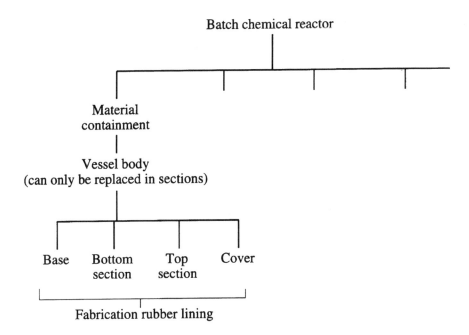

Figure 8.21. Maintainability diagram for vessel and lining

detected at an early stage it can be repaired in situ using a cold bonding process (see Appendix 2), a task which takes 2½ days, as compared with complete replacement (which takes ten days). Extensive deterioration can contaminate the product, causing a diminution in customer confidence in product quality.

A simple maintainability diagram for the vessel assembly and lining is shown in Figure 8.21. The cost of re-lining is high, both as regards material and labour; an in situ repair is by far the cheaper action. Thus, the repair versus replace decision is left (see Figure 8.8) until the lining is inspected (the inspection history also being used as the basis for guidelines for making this decision). For safety reasons, among other things, operate-to-failure is not an acceptable policy.

Thus the range of effective policies is as follows:

i. Fixed-time maintenance (vessel isolation, visual and tactile internal inspection then repair versus replace as necessary).
ii. Condition-based maintenance via
  (a) on-line fibre-optical or televisual inspection through the manhole (without isolation);
  (b) electro-chemical monitoring;
  (c) product sampling for contamination.

Table 8.6. Procedure for the vessel lining

Item	Timing	On-line or off-line	Frequency	Time and labour	Maintenance action	On-line or off-line	Frequency	Time and labour	Secondary action
Reaction vessel rubber lining	Visual inspection and touch	Off	Annually at agreed shutdown	3 days 2 fitters	Repair as necessary	Off Continuation of shutdown	Two yearly	1 day 2 men	Replace lining 8 days 4 fitters 2 men

Option ii(b) is only possible if the rubber has been penetrated, allowing a current to flow; this is not acceptable so the option is only available as a safety device to give warning of such penetration. Option ii(c) has been investigated but rejected because of the high rate of deterioration once contamination has begun; it also could be used as a shutdown warning but each batch of product would need to be sampled. The choice is therefore between option i and option ii(a).

There are clear advantages in on-line inspection because of the considerable time needed to isolate and wash the reactor before internal inspection and repair can be undertaken. On-line inspection, however, needs optical aids because it is impossible to see or touch enough of the lining from the manhole. Fibre-optical methods have been tried but are of limited use due to their small field of vision. Television did prove better but was considered insufficiently reliable without a tactile inspection back-up. On-line inspection is still being investigated but in the mean time fixed-time internal inspection — with repair or replace on-condition — remains the selected procedure (see Table 8.6). Using the limited data available the optimum interval between such inspections appears to be twelve months, but this can be reviewed as more data is collected.

2. The rotary joint of a paper machine

The dryer section of a large paper machine is made up of some 22 steam-heated rotating cylinders of the type shown in Figure 8.22. The paper is dried as it passes over each cylinder. The machine is operated continuously and its unavailability cost is high. Each cylinder has a rotating joint, the function of which is to allow steam to enter the cylinder and condensate to exit — while the cylinder is in motion.

Steam enters the joint (see Figures 8.22 and 8.23) via a flexible hose and passes through the joint on the outside of a syphon pipe. It then condenses in the cylinder, returns to the joint via the syphon, and leaves it through the condensate head, a flexible hose carrying it to sink. The rotating part of the joint is made up of a quick release mechanism, shank, spherical washer and

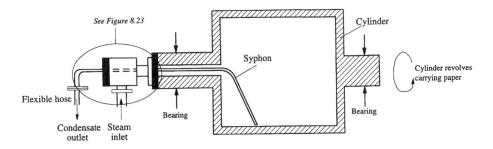

Figure 8.22. Rotary joint and heated cylinder assembly

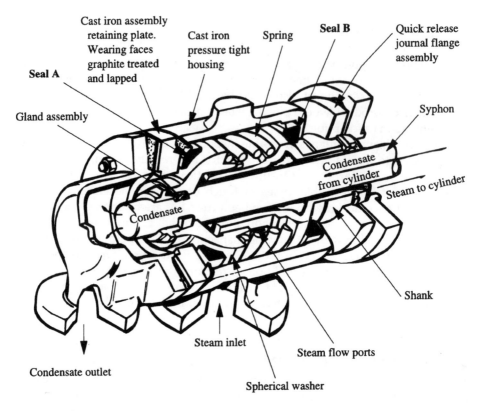

Figure 8.23. Rotary joint

gland assembly. During operation, seals A and B (see Figure 8.23) wear, but it has been found that seal A fails first, allowing steam to escape, damaging the paper and therefore precipitating machine shutdown. A spring provides the sealing force between the shank/seal-B and washer/seal-A interfaces and also promotes self-adjustment of the joint as the seals wear. Steam leaks often require the machine to be taken off-line. Replacement of the joint after failure or for seal inspection takes about two hours of off-line work. The time-to-failure of the joints ranges from six to twenty months, with a mean value of about thirteen months.

A simple maintainability diagram for the cylinder assembly is shown in Figure 8.24. The rotary joints are costly but repairable. It is much quicker to replace than to repair them in situ. Thus, bearing in mind the high cost of machine unavailability and the large number of rotary joints in use, the best maintenance action is *replace and recondition* (see Figure 8.7).

The effective maintenance policies are:

(a) *Fixed-time replacement.* Because of the poor statistical predictability and the high unavailability cost the interval between replacements might need to be as short as six months and would therefore carry

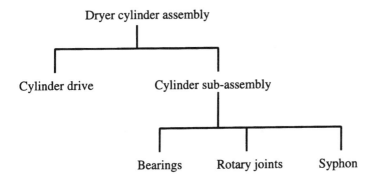

Figure 8.24. Maintainability diagram for dryer cylinder assembly

very high direct maintenance costs.
(b) *Operate-to-failure.* Would be far too expensive as regards maintenance costs (but should be considered).
(c) *On-line condition-based maintenance,* using a simple visual means of inspecting the wear of the joint seals. This has been proposed after careful consideration of Figure 8.23. It can be seen that the spring takes up wear by pushing the housing and condensate head away from the rotating part of the joint, the flexible hose on the steam inlet and condensate outlet allowing this movement. Therefore, the wear on the seals can be monitored by measuring the movement of the housing relative to the rotating part of the joint.

In this case option (c) is by far the most economic. It is outlined in detail in Table 8.7. Note that the timing of the off-line procedures must take into consideration the operational scheduling characteristics of the plant and unit.

3. *The roller element bearings of a paper machine*

The machine of the previous example contains many roller element bearings, including the two main ones on each of the heated cylinders (see Figure 8.22). It has been in use for some time and the available data suggests that the MTTF of the bearings is about five years, although some fail within months while others last many years. The lost-production cost of bearing failure is high.

The maintenance action is *replacement of the bearing* (see Figure 8.7). The effective policies to be considered are condition-based maintenance using shock pulse monitoring, and operate to failure. Operate to failure is unacceptable because of the high unavailability and repair costs. The economically appropriate procedure is condition-based maintenance based on monthly inspections and trend monitoring (see Table 8.8).

Table 8.7. Procedure for rotary joint

Item	Timing	On-line or off-line	Frequency	Time and labour	Maintenance action	On-line or off-line	Frequency	Time and labour	Secondary action
Rotary joint (22 off)	Visual condition checking	On	Monthly on-line maintenance routine for 22 joints	1 h for routine 1 fitter	Replace joints (about 1 per month)	Off	About once per month on agreed shutdown	2 h per joint (1 fitter)	Recondition

Table 8.8. Procedure for cylinder bearings

Item	Timing	On-line or off-line	Frequency	Time and labour	Maintenance action	On-line or off-line	Frequency	Time and labour	Secondary action
Bearings	SPM trend monitoring	On	Weekly as part of a running maintenance routine	4 h for routine (1 fitter)	Replace bearings (approx 2 per month)	Off	6 monthly at wire belt shop	2 h (1 fitter, 1 rigger)	None

4. *The wear plate of a chipping machine*

The chipping machine (Figure 8.25) is part of a series-structured continuously operated sawmill. The life of the mill is expected to be about nine years. Maintenance is normally undertaken during the two-week annual shutdown. Outside this window of opportunity production is lost if the mill is stopped. As regards the chipping machine the principal maintenance task is the replacement of the wear plate, unexpected failure of which results in ten hours lost production while it is replaced. If the production department is given advance notice of plate replacement only two hours of production are lost. Wear on the plate is due to abrasion by the wood chips. This cannot be seen from the mouth of the machine.

The costs involved are as follows:

Production loss per hour of downtime	£100
Labour, for replacement after failure	£350
Labour, for planned replacement or for removal for inspection	£100
Plate	£1000

The life expectancy of the wear plate is 18 ± 3 months (see Figure 8.26). It is known that wear progresses linearly with time, the rate being at worst about 2.5 mm/month. The management of the mill considers the present policy of ' operation to failure and replacement' to be too expensive. They have decided to investigate alternative procedures and have established the following additional information:

- Hard coating welding techniques to replace the material worn off the plate and return it to an 'as new' condition could be used every

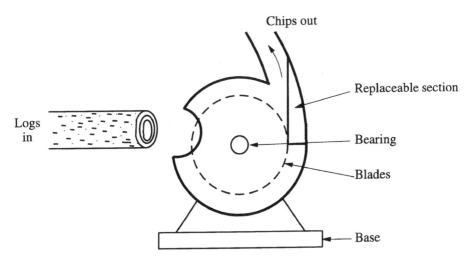

Figure 8.25. Schematic of chipping machine

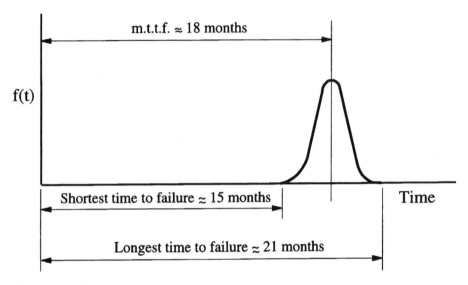

Figure 8.26. Distribution of wear plate failures

year, at a cost of £300. The repair takes an average of twelve hours.
- A re-designed wear plate with improved wear-resistant material costs £5000, but has a life in excess of nine years.

Both replacement and repair are possible maintenance actions and they need to be costed in conjunction with the effective policies.

The life of the wear plate is very predictable and fixed-time maintenance would be effective. The two most obvious timings for this are to do it every twelve months (in the annual production shutdown) or every fifteen months (which avoids failure but incurs a production loss of two hours).

What is not obvious is the method to be used to monitor the wear of the plate should condition-based maintenance be adopted. Ultrasonic techniques could be used, but a far less sophisticated procedure would be to drill one or more small-diameter holes (where the wear is likely to occur) to a depth of, say, 5 mm and to check monthly, with piano wire, to see whether the hole has been penetrated. This would give adequate warning of wear plate failure and the need for a replacement. If the wear rate was not known, trend monitoring could be facilitated by drilling a line of small-diameter holes of different depths.

An elementary life-cycle-cost comparison of the procedures, assuming an expected mill life of nine years, has been carried out (see Table 8.9). Because there are no major safety factors the aim is to minimize the sum of the direct and indirect costs. It can be seen that the most cost-effective procedure is to repair every twelve months, i.e. in the annual production shutdown. The details of the procedure are given in Table 8.10.

Table 8.9. Determination of best procedure for wear plate

Maintenance procedure	Average number of replacements over nine years	Expected production-loss cost (£)	Labour cost (£)	Materials cost (£)	Total cost (£)
Operate-to-failure and replace	6	6 x 10 x 100 = 6000	6 x 350 = 2100	6 x 1000 = 6000	14 100
Fixed-time replacement	8 (12 month period)	0	8 x 100 = 800	8 x 1000 = 8000	8800
	7 (15 month period)	6 x 2 x 100 = 1200	7 x 100 = 700	7 x 1000 = 7000	8900
Condition monitor and replace	6	6 x 2 x 100 = 1200	6 x 100 + 500 inspection cost = 1100	6 x 1000 = 6000	8300
Operate-to-failure and replace plus second line repair	6	6000	2100	1800	9900
Fixed-time repair (annually)	8	0	800	2400	3200
Re-design	0	—	—	—	5000 (Design cost)

Universal maintenance procedures

Ideally, the unit manufacturer should specify the maintenance procedure for each item, as a part of the unit life plan. Alternatively, manufacturers should describe the maintenance characteristics of their products so that the user may establish the best procedure. In practice, manufacturers supply limited information on procedures (and mainly for short-life simple items) and even less on maintenance characteristics. This is especially true for process plant. Various consultants now provide guidelines, or what they call 'Universal Maintenance Procedures' for many of the commonest items that can be found in industrial plant. An example of this is given in Table 8.11 and it can be seen that much useful information, for determining the best procedure, is provided[1].

Assembling the maintenance life plan for a unit

We have now reached the point where we need to consider the best way of assembling the various identified maintenance procedures into a complete maintenance life plan for the unit. To illustrate how this might be done, let us

Table 8.10. Procedure for chipping machine wear plate

Item	Timing	On-line or off-line	Frequency	Time and labour	Maintenance action	On-line or off-line	Frequency	Time and labour	Secondary action
Wear plate	Visual inspection against 5 mm limit touch	On	Monthly operator monitoring	Part of operator routine	Repair by welding	Off	12- monthly at mill shutdown	2 h 1 fitter 12 h 1 welder	None

look again at our example, the chemical reactor.

The determination of the procedure for the reactor lining was explained earlier (see Table 8.6). Employing a similar analysis procedures would need to be identified for the other reactor items (gearbox, motor, bearings, etc.) and listed in a similar way. It must be emphasized that the procedures and, in particular, their timing are decided in the light of the operational scheduling characteristics of the reactor (see Figure 8.27). This facilitates the selection of periodicities that are compatible with plant-wide scheduling.

An extract from the reactor life plan is shown in Table 8.12. The frequency of the major off-line *preventive work* is based on the agreed yearly shutdown of the reaction streams. The replacement of the lining and overhaul or replacement of the vessel are condition based. The minor preventive work is undertaken on-line (or can be carried out in windows of opportunity resulting from the pattern of production) and at intervals of one or more months.

Stand-by units and the life plan (see also Chapter 6)

The milling system of Figure 7.2 is an excellent example of the use of stand-by units to improve the reliability and maintainability of process plant. In this particular case Production required that any two of the three mills should be available continuously. The third mill could therefore be regarded as a stand-by unit (i.e. available in the event of one of the operating mills unexpectedly failing) other than when it was the one which was undergoing its scheduled major off-line maintenance (during which time there would be no stand-by, of course). In Chapter 6 we discussed how the reliability of such a system might be assessed. The following are some guidelines for its operation and maintenance.

- The system user-requirement must be clearly specified (see, for example, Figure 7.14).
- The life plan for each unit should be determined using the approach outlined in this chapter. When on stand-by a unit can be considered to have a hidden function, so its life plan should incorporate some form of proof testing, which might be inspection and checking by the operator.
- The units should be operated in a way which reduces the likelihood of:
  - (a) several running-units at a time being in a poor condition, i.e. of simultaneous failures creating a demand for replacements greater than the number of stand-bys — which, of course requires the operating histories (e.g. running times since last maintenance) to be noted;
  - (b) running-unit failures when a stand-by is in maintenance (e.g. the units that will be running throughout this period should be inspected at the start of it).

Table 8.11. Universal Maintenance Procedures; an example

Item	*Classification:* Brake: mechanical friction: operated by mechanical link mechanism: strongly corrosive surroundings; operated more than five times per shift; sudden break-down may cause fatal accident or damage exceeding £1500. (Treat separately: Lifter 3.)		*Universal maintenance item guide list*
Standard defects	*Defective components*	*Description of defect*	*MTTF*
	Steel parts	Corrosion	4 y
	Link bearings	Wear	4 y
	Nuts, other fastening elements	Loosening	Never
	Operated braking element, friction area	Wear	3 m
	Braked element, friction area	Wear, grooved	4 y
Universal Maintenance Procedures	*Description*		*Frequency*
	Activate. Measure time taken to stop (max $a =$ ____ s, 0.5 s + operating time of mechanism)		8 hourly
	Watch out for wear on braking element (thickness of lining min $b =$ ____ mm, 70% of new thickness); watch out for grooves on braked element (depth max $c = 0.5$mm); watch out for corrosion, lubricate $d =$ ____ (General Specification M.38)		Monthly
	Watch out for loose bolts; check play of link bearings (total free movement should not exceed 30% of normal lifter travel, i.e. max $e =$ ____); watch out for wear of braked element (wall thickness min $f =$ ____ mm, minimum 70% of original).		Yearly
Remarks	*Component breakdown:*		

Component breakdown:

1. *Mechanism*   2. *Friction elements*

1.1 Levers   2.1 Braking el. (Shoes, etc.)
1.2 Links and bearings   2.2 Braking el. (Drum, disc, etc.)
1.3 Bolts

                           3. *Lifting elements*
                               (not included)

                           4. *Closing elements*
                           4.1 Spring, weight, etc.

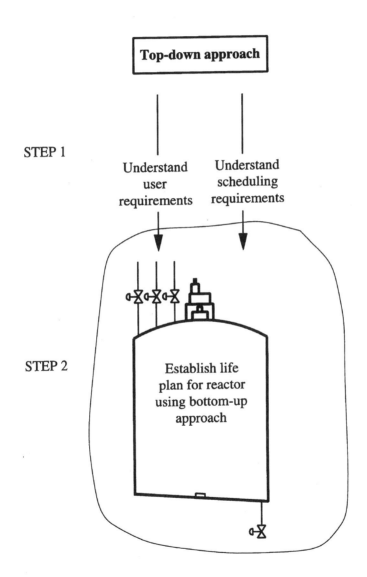

Figure 8.27. Reactor scheduling characteristics

Table 8.12 Extract from reactor life plan

Item	Timing	On-line or off-line	Frequency Y = Yearly M = Monthly	Time and labour	Initial maintenance action	On-line or off-line	Expected frequency	Time and labour	Action
Material Containment									
Rubber lining	Visual inspection and touch	Off	Y (at agreed shutdown)	3 days 2 fitters	In situ repair	Off	2 Y (WO)	1 day 2 trade assistants	Replace lining 8 days 4 fitters +2 trade assistants
Steel casing (pressure vessel)	Statutory inspection	Off	2 Y (at agreed shutdown)	2 days 2 fitters 1 inspector	In situ repair of jacket and joints as necessary	Off	4 Y (WO)	2 days 2 fitters	Replace vessel on condition
Agitation system AC Motor									
Bearings	SPM* trend monitoring	On	2 M	mins 1 inspector	Replace motor on condition	Off	4 Y (R)	1 h, 1 fitter 1 electrician	Recondition motor
Greasing	Fixed-time	On	2 M	mins 1 greaser	-	-	-	-	-
Gearbox									
Bearings	SPM trend monitoring	On	2 M	mins 1 inspector	Replace box on condition	Off	4 Y (R)	4 h 2 fitters 2 riggers	Recondition gearbox
Greasing	Fixed-time lubrication	On	2 M	mins 1 greaser	-	-	-	-	-
Gears	Lubrication, oil trend monitoring	On	2 M	mins 1 inspector	Replace box on condition	Off	?	4 h, 2 fitters 2 riggers	Recondition gearbox
Gland	Visual inspection	On	1 M	mins 1 inspector	Adjust gland	On	6 M	1 fitter mins	Repack gland on condition
Weigh vessel system									
Weighing mechanism	Visual condition checking	Off	6 M	1 h 1 inspector	Calibrate mechanism	Off	1 Y	30 min 1 inspector	Recondition mechanism on condition
Powder feeder	Visual condition	Off	6 M	1 h 1 inspector	Calibrate mechanism	Off	1 Y	30 min 1 inspector	Recondition mechanism on condition

Table 8.12. Extract from reactor life plan (Cont'd)

Item	Timing	On-line or off-line	Frequency Y = Yearly M = Monthly	Time and labour	Initial maintenance	On-line or off-line	Expected frequency	Time and labour	Action
Mat									
Hydraulic drive	Lubrication, oil trend monitoring	Off	2 M	mins 1 inspector	Replace drive unit	Off	3 Y	1 h 1 fitter	Recondition drive unit
Valves									
Type A (5 off)	Fixed-time	Off	2 Y	1 h 1 fitter (per valve)	Replace valve				Recondition valves (internal)
Type B (10 off)	Fixed-time	Off	5 Y	1 h 1 fitter (per valve)	Replace valve				Recondition valves (internal)
Type C (pressure relief)	Fixed-time	On	1 M	Inspection mins	Proof test				Replace and recondition as necessary

* Shock pulse monitoring   WO = Wear out   R = Random

References

1.   Davidson, J. and Hunsley, C., *The Reliability of Mechanical Systems*, 2nd edn, Mechanical Engineering Publications, IMechE, London, 1994.
2.   Grothus, H., *Universal Maintenance Procedures*, Plant Engineering Institute, D–427, Dosten 2, Wettring 4, Germany.

9

Determining the life plan and schedule — the top-down bottom-up approach

Introduction

In the previous chapter we explained how a life plan for a single unit of plant can be systematically determined. The next step is to consider how to incorporate this into a procedure for formulating the maintenance strategy for the whole plant, which may well be large and complex.

In accomplishing this the main tasks will be:

(a) formulating a *maintenance life plan* for each unit of plant (see Figure 9.1(a));

(b) formulating guidelines for setting up a plant-wide *preventive maintenance schedule* (see Figure 9.1(b));

(c) ensuring that the resulting workload can be properly resourced.

In theory, task (a) should be straightforward; for each unit, some form of life plan should be provided by its manufacturer. However, manufacturers' life plans vary from the excellent to the abysmal; at worst they may only be

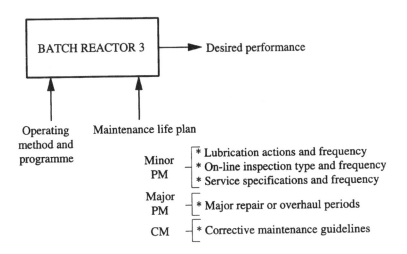

Figure 9.1(a). A typical unit and its maintenance life plan

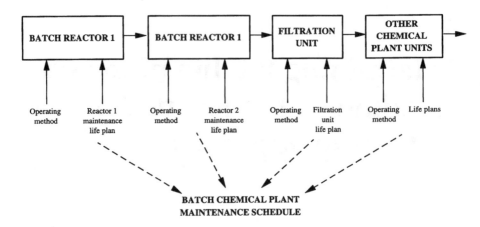

Figure 9.1(b). Assembling a maintenance schedule

lists of lubrication routines. In addition, in a large plant there may be hundreds of units each with a different manufacturer and in most cases the life plan will be written for general use, e.g. the recommended job frequencies will not be appropriate to a particular plant's scheduling characteristics.

The determination of a preventive maintenance schedule for the whole plant (see Figure 9.1(b)) is an inherently difficult task. The schedule is the sum of the multitude of maintenance procedures contained in the unit life plans and is impacted by many factors prominent among which are:

- the plant structure, e.g. the level of redundancy at plant stream, unit and item level;
- the capacity for interstage and final-product storage;
- the level of occurrence of unscheduled corrective maintenance;
- the incidence of planned outages, e.g. for catalyst changes or statutory safety work.

Perhaps the most important factor is the relationship between the product and its market. In some cases the demand for a product can be constant and stable — such as is placed upon a generator required to supply base load power to the distribution grid — while in others it may be variable and uncertain — in which case the maintenance schedule must be dynamic, i.e. responsive to production needs.

Resourcing the schedule can be difficult if the maintenance workload has major peaks, e.g. during petroleum refinery turnarounds, because the required contract labour may be of limited availability. Resources have to be provided not only for the preventive but also for corrective work, the latter often taking priority (leading inadvertently to changing from a preventive strategy to one based on operation-to-failure).

In short, the formulation of a maintenance strategy for a large plant is an involved and complex problem, the resolution of which requires a systematic, and pragmatic, approach. Such an approach — which the author calls *top-down*

Table 9.1. The top-down bottom-up approach

Step 1 *(Top-down)*	*Understand the plant structure and the characteristics* *of its operation*

(a) Construct a process flow diagram and establish a plant inventory.
(b) Understand the plant operating characteristics and the production policy.
(c) Establish the user requirements for the plant, plant sections and units. Rank the units in order of their importance (criticality) to output/safety/longevity.
(d) Understand the maintenance scheduling characteristics of the plant (using information from 1(a) and 1(b)).

Step 2
(Bottom-up)

Establish a maintenance life plan for each unit

(a) Identify the manufacturer's unit life plan (new plant) and/or the existing unit life plan and establish if they are likely to meet the requirements of 1(c) and 1(d) in a cost-effective way. If yes, record it as the unit life plan. If no, move to 2(b).
(b) Establish a new life plan or revise the existing life plan
   (i) Analyse the unit into its maintenance-causing items.
   (ii) Categorize the items according to their maintenance characteristics. Identify items with special functions.
   (iii) Determine the best procedure for each item. Deal with normal function items separately from 'special function' items. Special function items usually maintained via 'proof testing'.
   (iv) Assemble the maintenance procedures in the form of a unit life plan, (see Table 8.12). As far as possible the list should identify jobs with frequency, resource and method.
(c) Identify the need for spare parts and/or repairable items for each unit. Link such requirements to the overall spares inventory policy.

Step 3
(Bottom-up cont.)

Establish a maintenance schedule for the plant

(a) Prepare a plant listing of maintenance work by unit (based on the inventory of 1(a)).
(b) Establish the minor preventive schedules.
(c) Establish the major preventive schedules.
(d) Estimate from 3(b) and 3(c) the resource requirements for the scheduled workload. Forecast from experience the expected non-schedulable workload. Consider the effect that resourcing the workload might have on the maintenance schedule — in particular the shutdown schedule. Change the schedule as necessary.

bottom-up (TDBU) — is an integral part of Business Centred Maintenance. It is outlined in Table 9.1.

The top-down bottom-up (TDBU) approach

The iterative approach outlined in Table 9.1 encapsulates the *top-down* analysis of Chapter 7 and the *bottom-up* analysis of Chapter 8.

In Step 1 — in which the relative importance (to safety and economics) of the constituent units of the plant is established — those units are first identified and then a process flow diagram of the plant drawn up. *User requirements* are identified using the approach outlined in Chapter 7 (see Figures 7.13 and 7.14) and the *maintenance scheduling characteristics* of the plant studied.

In Step 2 maintenance procedures are identified and assembled into *unit life plans*. Much of the information derived during Step 1 is used for decision-making in Step 2.

In Step 3 the information derived during Step 1 , e.g. scheduling characteristics, is used in assembling the *plant preventive maintenance schedule*.

It must be emphasized that this approach should be regarded as a guideline, for assisting the maintenance department to formulate a new strategy or to revise an existing strategy. Because of the large differences, in structure and in operating characteristics, between one plant and another the approach might well need some modification in any particular case.

Step 1. Understanding the structure and characteristics of operation of the plant (the 'top-down' stage of the analysis)

(a) Construct process flow diagrams and draw up a plant inventory

Construct process flow diagrams such as those of Figures 7.1 (modelling the complete system) and 7.2 (modelling its sub-systems). The former should indicate plant production rates, raw material storage capacity; inter-stage storage and final product(s) storage capacities; the latter the way the units have been structured to perform process functions — for example, that two out of three mills are needed to operate to fully meet the requirements of the milling process (see Figure 9.2). The latter should identify unit production rates and product quality requirements, unit operating patterns and inter-process storage.

Use the process flow diagrams to make up an inventory of the plant at unit level, tying up with the documentation system and essentially a list of units, with a description and an identity code (see Figure 9.3).

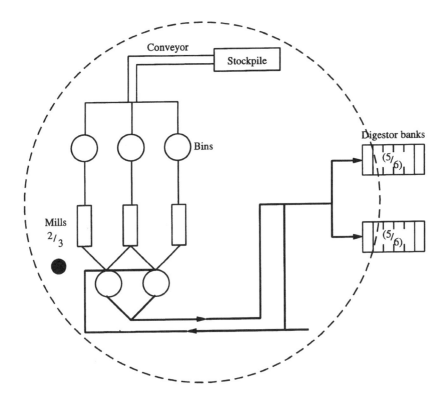

Figure 9.2. Process flow of milling system

(b) Understand the plant operating characteristics and production policy

Identify the production operating policy for the plant (arriving, through discussion and investigation, at a description which is not disputed by any of the parties concerned). This should include the process relationship between all the plants making up the operation (e.g. the mine, the ore processing and transportation plant, the refinery and the power station of Figure 7.1). In each case it is necessary to identify the plant operating pattern (shifts per day, days per week, weeks per year, with seasonal variations) and the expected product output, mix and quality.

Determine how other production factors — such as raw materials supply — and any external factors — such as safety regulations — influence the operating pattern of the plant and/or unit(s). Estimate the lost-production costs for the plant and for its units. Determine whether these are constant or variable (in power generation, for example, they might depend on the time of day; in agrochemical production on the time of year).

Batch chemical company,
made up of five plants.

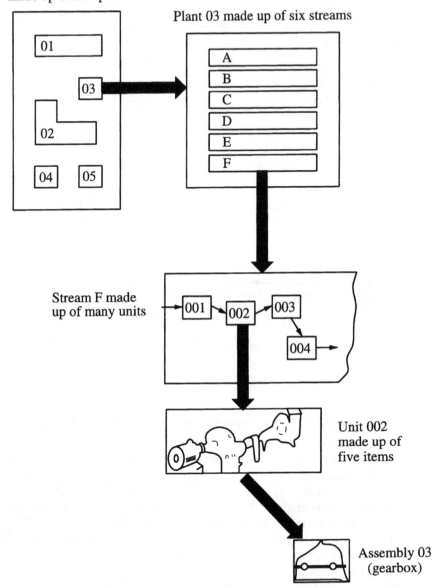

Figure 9.3. A simple hierarchical coding system

(c)　Identify the user requirement for each unit

Using the information gathered in (a) and (b) identify the plant user requirement, e.g. in the case of the refinery this was a 95 per cent availability at a stated product quality. This was translated at unit level as a requirement for two out of three mills to be operating — to achieve 100 per cent availability of supply of milled product to a stated standard. User requirements are identified via the approach outlined in Figures 7.13 and 7.14.

It is often useful as a part of this step to rank the units in order of importance, or *criticality*, i.e. according to the impact of their failure on production and safety, e.g.

Level 1 — causes a serious loss of safety,
Level 2 — causes an immediate loss of production,
Level 3 — does not have an immediate effect on production or
　　　　　　safety.

The information derived at this stage is essential for formulating, in Step 2, the unit life plans.

(d)　Understand the maintenance scheduling characteristics of the plant (using information from (a) and (b))

(i)　Identify the opportunities, the production windows, for off-line maintenance. These may result from seasonal, monthly, weekly or daily variations in demand for the product, e.g. in the case of the refinery there are no plant windows — the operation is continuous. Some of these windows may be well defined in terms of frequency and duration, e.g. those arising from statutory pressure vessel inspection. In other cases they may occur with much less certainty, e.g. due to fluctuating demand for the product. This is one of the most important characteristics influencing *opportunity scheduling*, such as may be desirable for a power station on two-shift operation tending to generate windows of up to one week's duration which occur with random incidence (see Case Study 1, Chapter 11).

Production-related windows can arise:

- for a unit, or a group of units, because of production scheduling, e.g. in a multi-product plant where a given product mix does not require a particular production line;
- for the plant, or for units, because of production changes, e.g. catalyst changes;
- for a unit, or a group, because of the availability of redundant or stand-by equipment (see, for example, Figure 9.2 which indicates such windows at the mill level);
- for a unit, or a group, because of the availability of interstage storage and excess capacity;
- for the whole plant, or for units, because of statutory safety work.

(ii) Identify 'domino' situations, where the effects of off-line maintenance on a unit propagate along a batch process. Inter-stage storage prevents the whole line coming off and spreads the maintenance for the line over a longer period, i.e. it smooths peaks.

(iii) Identify 'process chains' where, in order to maintain a single unit, a whole process involving many units needs to be taken off-line. This either causes maintenance resource peaks or excessive planned downtime for maintenance.

Information from this step is essential to setting up the maintenance schedule (part of Step 3).

Step 2.   Establishing a life plan for each unit of plant — the 'bottom-up' analysis

This step uses the information from Step 1 — in particular from parts 1(c) and 1(d) — in order to decide on the amount of preventive maintenance that can be justified for each unit, see Figure 9.4. Part 1(c) provides essential

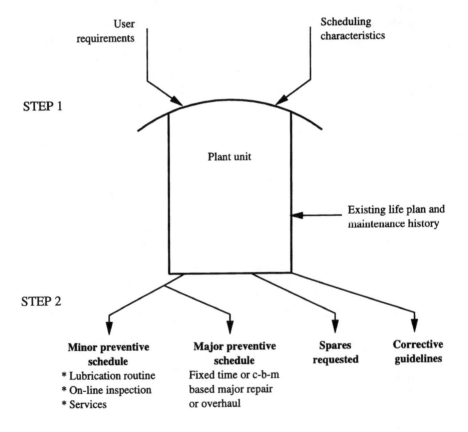

Figure 9.4.   The relationship between Step 1 and Step 2

economic and safety information, Step 1(d) the essential scheduling information, e.g. in the case of the milling system the frequency of off-line maintenance can be decided at unit level — taking a mill off line does not affect other units. In other cases part 1(d) will have identified the incidence and duration of windows for individual units.

Step 2 now proceeds as follows:

(a)   Identify the manufacturer's — or the existing — life plan

Identify the life plan recommended by the manufacturer (or the existing life plan) and establish whether it is likely to meet the *user requirements* and can be carried out within the *scheduling opportunities*.

The less important units of plant (which could well be the majority) might only require a minimum level of maintenance (the minor, scheduled, preventive work), a 'wait and see' policy. The schedule will mostly be provided by the manufacturer, or be already in existence. For the highly critical units of plant, those deemed important as regards economics, safety, or longevity, the more detailed analysis of Step 2(b) may be necessitated.

(b)   Formulate a new life plan or revise the existing life plan

The principles associated with this step were outlined in Chapter 8, using the example of a chemical reactor. The same example will therefore be used here.

(i)   Analyse the units into their maintenance causing items. An example of this is outlined in Figure 9.5. A more pragmatic approach might be to identify the maintenance causing items via the manufacturer's manuals or drawings.

(ii)   Categorize the items as shown in Figure 9.6. In particular identify the special items which are handled separately via some form of proof testing.

(iii)   Determine the best maintenance procedure for each identified item. The best procedure for the vessel rubber lining, for example, is shown in Table 9.2. These procedures will be the building bricks of the life plan.

(iv)   Assemble the procedures into a unit life plan. One way of recording a unit life plan would be in the form of the listing shown in Table 9.3. This is a statement of the maintenance work required on the reactor over its expected lifetime. The list should also be in a form suitable for assimilation into the overall *plant maintenance schedule* so it will be necessary to adopt standardized frequencies (based on calendar or running time) for maintenance work — in particular for off-line work which should link with the information coming through from Step 1(d). Referring to the reactor example, the following actions might also be helpful:

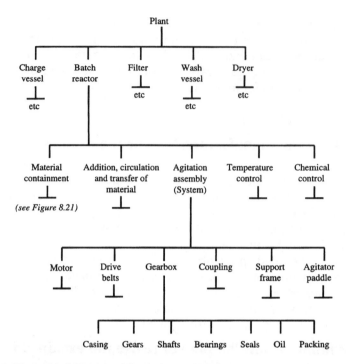

Figure 9.5. Hierarchical division of batch chemical plant

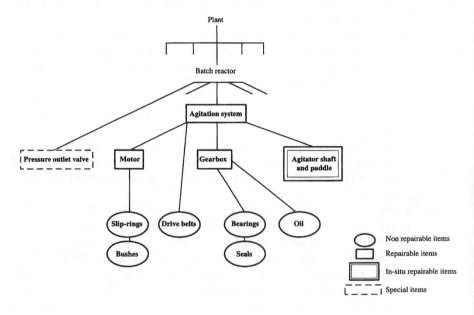

Figure 9.6. Maintainability diagram for batch reactor

Table 9.2. Procedure for the vessel lining

Item	Timing	On-line or off-line	Frequency	Time and labour	Maintenance action	On-line or off-line	Frequency	Time and labour	Secondary action
Reaction vessel rubber lining	Visual inspection and touch	Off	Annually at agreed shutdown	3 days 2 fitters	Repair as necessary	Off Continuation of shutdown	Two-yearly	1 day 2 men 8 days	Replace lining 4 fitters 2 men

- Divide the procedures listed in Table 9.3 into on-line and off-line. On-line procedures, by definition, are independent of production, and should be considered for scheduling on a plant or area wide basis.
- As far as possible, group the off-line procedures (including inspection-based ones) by trade and periodicity into 'jobs' and formulate the instructions and times for them, e.g.
  - Mech. A Service. Monthly. 5 h, Job code 125M
  - Elect. B Service, Three-monthly (inspection-based). 8 hrs, Job code 127E
  - etc.

For convenience of work planning a 'job' can be regarded as any maintenance task taking, say, less than 24 hours (three shifts) and needing no more than a few tradesmen. A job can be handled on a single work order card. Work taking longer than 24 hours and/or requiring many tradesmen can be classified as a major repair, reconditioning, or overhaul and can be made up of a number of 'jobs' — perhaps arranged into a standard package with bar-chart.

It will also be necessary to identify the jobs which might occur as a result of operating critical items to failure, e.g. while such an item, a gearbox say, is being replaced, the other parts of the associated unit or, for that matter, of any part of the plant which might be stopped by the gearbox failure.

The unit life plan should be reviewed periodically in the light of maintenance cost and reliability performance.

(c)   Identify the need for spare parts and repairable items

The analysis of Step 2(b) identifies the need for spare parts and for reconditioning. Such information can also come direct from the manufacturer's manual. In general, the maintenance engineer decides on what spare parts are to be held and on their specifications. The spares inventory policy is a function not only of demand for the various parts but also of such factors as the opportunity for interchanging parts. In general this task is the responsibility of the 'stock controller'.

The factors influencing reconditioning policy are also complex. For example, deciding whether to recondition or to replace with new may be based on a type of life-cycle analysis of the alternatives. Once again, an influencing factor is the rate of demand for an item and therefore it will be necessary to determine the number of identical items on site, see Figure 8.9. Using this information the design of the reconditioning cycle should address such aspects as:

- the determination of the reconditioning facilities needed,
- the determination of the manpower needed,
- the determination of the spare parts needed,
- the total number of floating items,
- the inventory policy for the reconditioned parts,
- the logistics of moving reconditioned parts around the cycle.

Table 9.3. Extract from reactor life plan

Item	Timing	On-line or off-line	Frequency Y = Yearly M = Monthly	Time and labour	Initial maintenance	On-line or off-line	Expected frequency	Time and labour	Action
Material Containment									
Rubber lining	Visual inspection and touch	Off	Y (at agreed shutdown)	3 days 2 fitters	In situ repair	Off	2 Y (WO)	1 day 2 trade assistants	Replace lining 8 days 4 fitters + 2 trade assistants
Steel casing (pressure vessel)	Statutory inspection	Off	2 Y (at agreed shutdown)	2 days 2 fitters 1 inspector	In situ repair of jacket and joints as necessary	Off	4 Y (WO)	2 days 2 fitters	Replace vessel on condition
Agitation system									
AC Motor									
Bearings	SPM* trend monitoring	On	2 M	mins 1 inspector	Replace motor on condition	Off	4 Y (R) 1 electrician	1 h, 1 fitter motor	Recondition
Greasing	Fixed-time	On	2 M	mins 1 greaser	–	–	–	–	–
Gearbox									
Bearings	SPM trend monitoring	On	2 M	mins 1 inspector	Replace box on condition	Off	4 Y (R)	4 h 2 fitters 2 riggers	Recondition gearbox
Greasing	Fixed-time lubrication	On	2 M	mins 1 greaser	–	–	–	–	–
Gears	Lubrication, oil trend monitoring	On	2 M	mins 1 inspector	Replace box on condition	Off	?	4 h, 2 fitters 2 riggers 1 fitter	Recondition gearbox
Gland	Visual inspection	On	1 M	mins 1 inspector	Adjust gland	On	6 M	mins	Repack gland on condition
Weigh vessel system									
Weighing mechanism	Visual condition checking	Off	6 M	1 h 1 inspector	Calibrate mechanism	Off	1 Y	30 min 1 inspector	Recondition mechanism on condition
Powder feeder	Visual condition	Off	6 M	1 h 1 inspector	Calibrate mechanism	Off	1 Y	30 min 1 inspector	Recondition mechanism on condition
Hydraulic drive	Lubrication, oil trend monitoring	Off	2 M	mins 1 inspector	Replace drive unit	Off	3 Y	1 h 1 fitter	Recondition drive unit
Valves									
Type A (5 off)	Fixed-time	Off	2 Y	1 h 1 fitter (per valve)	Replace valve				Recondition valves (internal)
Type B (10 off)	Fixed-time	Off	5 Y	1 h 1 fitter (per valve)	Replace valve				Recondition valves (internal)
Type C (pressure relief)	Fixed-time	On	1 M	1 h inspection mins	Proof test				Replace and recon- dition as necessary

* Shock pulse monitoring   WO = Wear out   R = Random

Step 3. Establishing a preventive maintenance schedule for the plant — 'putting it all together'

This step is concerned with deciding on the best way to schedule the hundreds (perhaps thousands) of individual jobs identified in the unit life plans, taking into consideration the maintenance scheduling characteristics identified in Step 1(d), i.e. the effect that off-line work might have on production and on maintenance resources. The scheduling procedure is shown in Figure 9.7 (see also Table 9.1).

(a)   Prepare a listing, by unit, of maintenance work for the plant

Prepare a list (the Main List) of all the maintenance work identified for every unit of plant, arranging the list in the order of process flow. This should tie up with the Plant Inventory of Step 1(a) (see also Figure 9.3).

(b)   Formulate the minor preventive maintenance schedules

This is made up of the on-line work and minor off-line preventive work.

(i)   Extract the on-line work (mainly inspection procedures) and group the jobs according to trade, geographical area or plant type, and frequency. Prepare job instructions for these routines. Such work can be scheduled independently of production and is made up of:

- operator monitoring routines,
- tradeforce inspection routines (line patrolling),
- instrument-based routines (undertaken by specialist internal or contract resources).

Such work is particularly important because it often leads to the identification of the need for major off-line work.

(ii)   The remainder of the work in this category is made up of simple off-line inspection or lubrication and the replacement of 'simple items', i.e. routine services. It is undertaken frequently and is of short duration and can almost always be fitted into production windows. It is essential that such work is carried out because it is directed at controlling the reliability of the complex items.

(c)   Formulate the major preventive maintenance schedules

This is a complex problem the solution of which will be different for different plants. The following pragmatic approach attempts to identify the key points of scheduling.

(i)   Where the required frequency of occurrence and duration of the work are less than those of the expected windows (see, for example, the machine shop case, Figure 9.8, or the refinery example, Figure 7.2).

This is by far the easiest situation for scheduling. The off-line work can be scheduled at unit level and spread out in order to smooth the

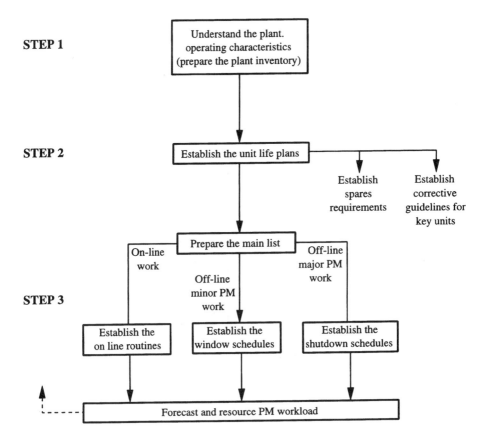

STEP 1 — Understand the plant. operating characteristics (prepare the plant inventory)

STEP 2 — Establish the unit life plans — Establish spares requirements — Establish corrective guidelines for key units

STEP 3 — Prepare the main list — On-line work — Off-line minor PM work — Off-line major PM work — Establish the on line routines — Establish the window schedules — Establish the shutdown schedules — Forecast and resource PM workload

Figure 9.7. Step 3 — the scheduling procedure

workload. This in turn makes for easier work planning and provides the opportunity for better resource utilization.

(ii) Where the required frequency of occurrence and duration of off-line work are greater than those of the expected windows (see, for example, the base load power station case, Figure 9.9).

Identify and schedule the work from the main list that can be carried out in the maintenance windows (in this case windows occurring because of the presence of standby or redundant units, e.g. as with most pulverizing mills). The remaining off-line work can only be carried out with the plant shut down.* The most straightforward situation is where most of the off-line jobs are time based. The time between plant shutdowns can be based on the shortest unit running-period (in our example the time between statutory safety inspections of the boiler). As far as possible, other jobs are fitted into this period and into multiples of it. Scheduling the plant shutdown on a time basis has considerable advantages for organizing the labour to match large work peaks.

There are numerous variations that can complicate this situation. For example, it is often possible to extend the shortest running period by applying condition-based maintenance to the critical items. However, this might shorten the planning lead time. A major shutdown is often chosen to coincide with, and be an extension of, a major window.

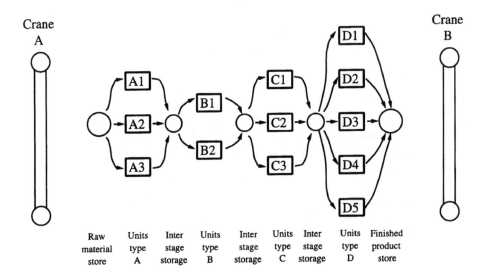

Plant structure:
Units in parallel/series structure with spare capacity at unit level plus interstage storage. Many unit windows

Plant operating pattern:
Single shift five day week (sales limited). Many plant windows.

Planned maintenance can be scheduled at unit level, smoothing the work load over a long period

Figure 9.8. Scheduling characteristics of a machine shop

(iii)   Opportunity scheduling
Because of the uncertainty associated with both the frequency and
duration of windows (and sometimes of the incidence of the off-line
work) it is inevitable that the scheduling of the off-line work will
involve a considerable level of opportunity-taking, e.g. taking
advantage of failure occurrences, or of unexpected windows, to carry
out planned preventive and/or corrective maintenance. Perhaps
the most difficult scheduling task is presented by plants which
generate a considerable level of major off-line work due to randomly
occurring failure — despite the application of preventive maintenance.
In such situations opportunity scheduling is perhaps the most
important policy. The more recent computerized work planning and
scheduling packages greatly facilitate this.

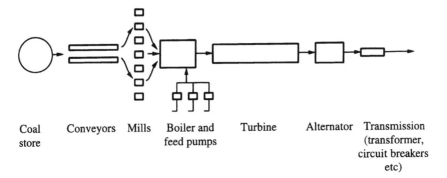

| Coal
store | Conveyors | Mills | Boiler and
feed pumps | Turbine | Alternator | Transmission
(transformer,
circuit breakers
etc) |

Plant structure:
Units in parallel/series structure with spare capacity
on parallel sections e.g. mills, feed pumps.
No interstage storage.

Plant operating pattern:
Continuous except for maintenance (production limited)
-no plant windows. Back up plant to cover
but much less efficient.

Planned maintenance scheduled by category
i) on-line inspection schedule at item level
(since independent operation) into a smooth
work load over long period.

ii) Window schedule at unit level (e.g. mills)
into a smooth work load over long period

iii) Shutdown schedule at plant level based
on safety inspection of boilers (three years)
Workload dominated by shutdown peaks
(up to 1300 men)

Figure 9.9. Scheduling characteristics of a power station

In this case 'plant' means a boiler-turbine-alternator stream or 'set'.

(d)   Resourcing the workload

The ability of the organization to resource the maintenance workload has been surveyed in the discussion of Steps 3(a) to 3(c). The principal factors affecting this are the planning lead time, the size of resource peaks and the availability of contract labour. Thus, where the resources are unable (or cannot be afforded) to meet the major outage demand the schedule would need to be revised. In the case of a power station, for example, this would depend on the ability to use contract labour. If that was inhibited by industrial relations factors the shutdown schedule would need modification.

The larger the work peaks the longer needs to be the planning lead time. In the case of a large power station, planning a shut-down takes twelve months — so fixed time shutdowns are invariably used.

TDBU application

The TDBU approach will now be illustrated using the batch chemical plant referred to in Chapters 4 and 8 as an example.

Step 1.   Understanding the characteristics of operation

(a)   Constructing process flow diagrams

The process flow of the plant operation is shown in Figure 4.3 and of a reaction stream in Figure 4.4.

(b)   Understanding the production policy

The plant, which manufactures a wide range of similar organic chemicals, comprises six reaction streams, three making soluble products and three making insoluble ones. Each stream is dedicated to a specific range of chemicals and the streams are not interchangeable. The finishing streams are also divided into those processing soluble and those processing insoluble products and are only interchangeable within these groupings. A plant path (as indicated in Figure 4.3 by the bold line) is a complete path through the plant from reaction process to packaging. A typical reaction stream, indicating the inter-relationship of the units, is shown in Figure 4.4. The reaction process operates on a 48-hour batch cycle while the finishing process operates on a 168-hour semi-continuous cycle. A number of reaction cycles have to be completed before enough chemical is stored to allow the finishing process to begin.

In addition to the main product flows, the plant is supported by a full range of chemical and engineering services, i.e. primary and intermediate chemical supplies, salt, flake ice (not shown). The reaction streams are computer controlled and the rest of the plant is remotely controlled so the plant can be operated by a small production staff.

The production plan is complex because of the many products manufactured, but there is a balance throughout the plant which means there is little or no spare capacity in the finishing streams if all the reaction streams are in use. At present the plant is production limited.

(c)   Determining the user requirement

Up until now an average availability level for the whole plant of 92 per cent has been achieved. Because of the demand for the product the maintenance objective for the next two years is to increase this figure to 96 per cent at no extra maintenance cost. The plant is about ten years old and has an expected life of thirty years or more, given appropriate life extension work. The plant can be regarded as hazardous because of its employment of corrosive chemicals at high temperature and pressure. The company is safety conscious and has recently introduced Du Pont safety procedures and standards.

The user requirements for the individual units are determined as outlined in Figure 7.13. This will be illustrated using Reaction Unit 3 as an example. This is a production-critical unit from which Production want 100 per cent reliability of operation over its 51-week operating period (i.e. 98 per cent annual availability) and high product quality (i.e. conforming to written quality standards). In addition they have stated that the reactor should be maintained in such a way that it will operate safely for its expected life.

(d)   Understanding the scheduling characteristics

The plant is operated on a full-time basis, i.e. 168 hours per week. During the Christmas week, however, the plant is not used by Production but the chemicals are held in process to minimize the effect of the holiday loss. There are no windows for long-term maintenance scheduling. There is an agreed maintenance shutdown of one week per plant path. Closer investigation shows that over any monthly period all the units in a plant path become available for maintenance, at short notice, for periods of between two and eight hours. These windows arise randomly because of the batch nature of the process and the various washing-out procedures that are required between different products. They can be used for small off-line jobs if there is good communication between production and maintenance.

It will be instructive at this point to consider how the batch nature of the process influences the scheduling of the major off-line maintenance. Because of the short cycle time (48 hours) of the reaction process each reaction stream has to be considered as a whole when major maintenance work needs to be carried out. In other words, if a single reaction unit is taken off-line for one day, then the whole reaction stream has to be taken off-line. This is not the case with the units in the finishing stream because the cycle times are much longer. Thus, if a reaction stream is taken off-line for, say, three days, this window of maintenance opportunity moves down the finishing stream unit by unit, the *knock-on* principle. The interchangeability between finishing streams makes the scheduling of maintenance in this window a straight-forward exercise.

Because there is a little interchangeability between reaction streams, and no spare capacity in a normal sales market, failure of the plant means high downtime cost.

Step 2.   *Establishing a maintenance life plan for each unit*

As above, Reaction Unit 3 will be taken as the illustrative example so some information about its construction is now needed.

Material containment
1000 gal mild-steel rubber-lined vessel built in ring sections.
Agitation system
DC variable-speed motor, worm reduction gear box, mild-steel rubber-lined paddle agitator.
Weigh vessel system
Mild-steel rubber-lined vessel (300 gal) — not shown, weighing mechanism, weigh scale instrumentation.
Pumping system 1 (recirculation and filter press feed)
DC variable speed motor, mono-pump, pump protection instrumentation, mild-steel rubber-lined pipework, GRP/PVC line pipework, valves and fittings.
Pumping system 2 (filter press feed)
As for pumping system 1.
Instrumentation and controls
DP cell (level), thermocouple in tantalum-clad pocket, steam injection posts (temperature), pH probe (chemical).

The vessels are considered as pressurized because of the steam injection and are subject to a two-yearly pressure vessel inspection.

Considerable information is available on the failure and deterioration characteristics of the pumps, motors, gearboxes, etc., but this is not the case with the rubber lining, for which the most that can be said is that it has a life which can be as little as three and as large as nine years, with a mean value of six years, and that the only method of monitoring its deterioration is by a combination of visual and tactile inspection (see Chapter 8, Example 1). The time to onset, and the subsequent rate of deterioration, are also uncertain but it is known that the rate is slow and the lead time to eventual failure exceeds one year. A lining inspection and repair takes 2½ days and a replacement ten days.

(a) Determining the existing maintenance life plan

The existing life plan was not documented; it was based on performing the essential lubrication on-line and undertaking major maintenance in the seven-day annual shutdown, when the main job was the inspection and repair of the rubber lining. Apart from this the work was carried out on an *ad hoc* basis and there had been a considerable level of corrective work — some of which had caused unavailability. The current reactor availability is around 90 per cent and many of the failures cause considerable disruption. It has also been noted that the condition of the reactor is below specification. Clearly, the existing life plan of the reactor is in need of radical review.

(b) Revising the existing life plan

(i) The reactor is analysed into its maintenance causing items (see, for example, Figure 9.5).

(ii) The items are categorized and those with special functions identified (see, for example, Figure 9.6). The pressure relief valve, for instance, has a special function and is proof-tested periodically.

(iii) The best maintenance procedure for each item is determined (see, for example, the analysis, in Table 9.2, for the rubber lining).

(iv) The procedures are assembled into a unit life plan (see Table 9.3).

(c) Identifying the need for spare parts

For this, we extend analyses such as that of Table 9.3, i.e. repair and/or replace decisions lead to the need for spare items or components. For example, in the case of the AC motor of the agitation assembly it is necessary to hold the complete motor. The spare components would be held only if the motor is to be reconditioned in-house.

Step 3. Formulating a preventive maintenance schedule (see Figure 9.7)

(a) **A 'main list' is created** (i.e. of the identified maintenance procedures for the whole plant, categorized according to the section, area, stream and unit to which they are applied) e.g.

```
REACTION STREAM F
03F001    REACTION UNIT 1
03F002    REACTION UNIT 2
03F003    REACTION UNIT 3
03F004    PREPARATION UNIT
```

03F005   FILTRATION UNIT 1
03F006   FILTRATION UNIT 2
03F007   DISPERSION UNIT
etc.

(b)   The on-line routines are determined

(i)   The on-line routines can be extracted from the main list and classified by trade and frequency. In the case of the reactor this would be as follows. (M = monthly; Y = yearly)

Fitter		*Inspector*		*Greaser*	
M	visual inspection and adjustment of agitation system gland.	M	Proof test Type C pressure relief valve.	2 M	AC resistor bearings.
		2 M	SPM AC motor bearings. SPM gearbox bearings. Lub monitor gearbox. Lub monitor hyd drive.		Gearbox bearings.

This approach can be used to assemble schedules of inspection and lubrication routines.

(ii)   Constructing the schedule of 'window maintenance'.
In this plant, window maintenance jobs can be classified as those that can be carried out in the production windows (i.e. jobs of less than two hours duration) and those that do not require the plant to be taken off line.

The procedures that can be carried out in windows of opportunity are also listed by unit and are grouped by trade and frequency, e.g.

03F001   Reaction Unit 1
6M Inspection. Inspector, 2.5 h.
* Visually inspect weighing mechanism and adjust and calibrate as necessary.
* Visually inspect powder feeder and adjust and calibrate as necessary.
  etc.
2Y Valve replacement. Fitter, 5 h.
* Replace type A valves (5 off).

The identified jobs are then scheduled for each trade and for a complete year. A bar chart as in Figure 9.10 is often used to assist this last process. Where there is a definite production window the job can be scheduled to the day. If the timing of the window is not known with certainty the job can be scheduled to the nearest week and fitted in when the window occurs. Off-line window work that is triggered by the output from condition monitoring routines, e.g. gearbox replacement, is not scheduled until the need has been thus indicated.

(c) The major preventive maintenance schedules are established

An extract from the maintenance shutdown plan for a reactor stream is shown in Table 9.4. The initial running time is based on the shutdown work for Reaction Unit 3 and the preparation unit. In subsequent years the longer term maintenance will be fitted into this shutdown, giving the following major shutdown schedule:

Yearly	Shutdown list A (reaction and preparation unit)
Two-yearly	Shutdown list B (list A plus lining repair, plus statutory inspection)
Four-yearly	Shutdown list C (list B plus steel casing repair)

Other work is fitted into the shutdown as necessary.

The yearly shutdown of the reaction stream also determines the plant path shutdown, which starts with the reaction stream shut-down (typically of four days duration). Work on the remaining plant sections and units is carried out as the window is 'knocked down' the path.

The plant comprises six 'plant paths' and in order to spread the load evenly over the year (to make the best use of in-house labour) there should be a major shutdown of a plant path every two months; the flexibility of the finishing stream aids such scheduling.

Comments

Is the new strategy outlined above an improvement over the former one? Will the availability, product quality, and equipment condition improve at no increase in resource cost? This is best answered by reference to the reactor and reaction stream examples.

The maintenance strategy has moved away from *annual shutdown with ad hoc planning* to *condition-based* founded on the following three inter-related preventive schedules:

(i)   on-line lubrications and inspections,
(ii)   services and minor maintenance undertaken in production windows,
(iii) a major plant shutdown.

Although the expected level of improvement is not easily quantified it is clear that the emphasis on condition-based maintenance should both minimize unexpected failures and improve the efficiency of shutdown planning. What should result from the new strategy is a movement from situation (a) of Figure 9.11 to situation (b).

STREAM No.	UNIT	1	2	3	4	5	6	7	8	9	10	11	12	13	14	15	etc
REACTION STREAM 1	Reaction unit 1	6M															
	Reaction unit 2		6M														
	Preparation unit			6M													
	Reaction unit 3				6M												
	Filtration unit 1					6M											
	Filtration unit 1						6M										
	Disperser unit 1							6M									
REACTION STREAM 2	etc	6M															
			6M														
				6M													
					6M												
						6M											
							6M										
								6M									

WEEK No →

Figure 9.10. Extract from reaction streams preventive maintenance schedule

Table 9.4. Extract from the shutdown plan for a reaction stream

Unit	Item	Timing	On-line or off-line	Frequency (Y = yearly)	Time and labour	Initial maintenance action	On-line or off-line	Expected frequency	Time and labour	Secondary maintenance action
03F003	Rubber lining	Fixed time (visual inspection and touch)	Off	Y (at agreed shutdown*)	3 days 2 fitters	Repair lining as necessary	Off	2 Y	1 additional day (4 days total) 2 fitters	Replace in situ on condition
03F004	Rubber lining	Fixed time (visual inspection and touch)	Off	Y (at agreed shutdown of unit 3)	3 days 2 fitters	Repair lining as necessary	Off	2 Y	1 day 2 fitters	Replace in situ on condition
03F001	Steel casing	Statutory inspection	Off	2 Y (at agreed shutdown of unit 3)	2 days 2 fitters 1 inspector	Repair jacket and joints as necessary	Off	4 Y	2 days 2 fitters	Replace vessel on condition
03F002	Steel casing	Statutory inspection	Off	2 Y (at agreed shutdown of unit 3)	2 days 2 fitters 1 inspector	Repair jacket and joints as necessary	Off	4 Y	2 days 2 fitters	Replace vessel on condition

*Schedule other maintenance work (window and deferred) into this shutdown as convenient

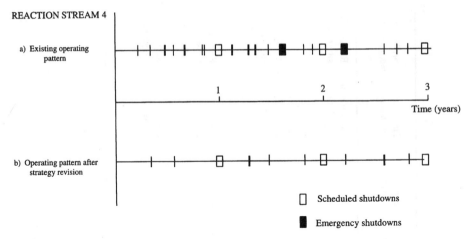

Figure 9.11. Improvement in operating pattern

It should also be emphasized that documenting the life plans as in Table 9.3 greatly facilitates the application of opportunity maintenance. This is especially so if the documentation is computerized. In the advent of an unexpected maintenance window (or a failure of some other part of the reaction stream) the maintenance schedules can be used to extract the maintenance jobs for the reactor that are outstanding at that time.

Using the TDBU approach

To date, perhaps the most successful application of this was to an up-rated and modernized blast furnace which was part of an integrated steelworks. The main lessons learned were as below.

- Such a project is best initiated and co-ordinated by a small inter-disciplinary project team with a project leader.
- Step 1 must involve the senior levels of production and maintenance management. Perhaps the most important part of this is 1(c) — Determining the user requirements. There is a sense in which the result of this step should be regarded as a *contract* between maintenance and production.
- The bottom-up phase — in particular Step 2 (Formulating a life plan for each unit) — must involve maintenance supervision, tradesmen and plant operators who are concerned with the particular units of plant under scrutiny. This is essential for benefiting from 'local knowledge' and for promoting future commitment to the life plan. It is important that maintenance supervision (and, to a lesser extent, tradesmen and operators) are regularly updated on the principles, concepts and techniques of preventive maintenance.

10
Controlling plant reliability

The previous chapters have dealt with the task of formulating the life plan for each of the units that make up a plant (see Figure 9.1(a)) and a maintenance schedule for the plant as a whole (see Figure 9.1(b)). Little has been said, however, about reviewing the performance of the units to determine whether the life plan is effective, i.e. whether the life plan is providing the desired output as regards all its various aspects — quantity, quality, continuity, etc. This — *the control of maintenance effectiveness* (see Figure 7.11) — is probably the most important maintenance control activity*. Once again, the alumina refinery will serve as the vehicle for explaining it (see Figure 7.2).

Figure 10.1, which outlines the mechanisms for controlling the effectiveness of one of the refinery units, illustrates the classic ideas of *reactive* control — using the feedback of operational and maintenance data — and also highlights *pro-active* control via the feedforward of ideas for reliability and maintenance improvement.

Reactive control of plant reliability

The requirements of the systems are to:

(a) monitor the output parameters of each unit, e.g. reliability (mttf), maintainability (mttr), plant condition, etc., and some of the input conditions, e.g. whether the unit life plan is being carried out to specification and at anticipated cost;

(b) determine the root cause of any failure (note that a control system for this must encompass several departments because the cause could originate in Production (maloperation), in Engineering (poor design) or in Maintenance);

(c) prescribe the necessary corrective action.

At refinery level, control can be envisaged as in Figure 10.2, i.e. each unit having its own control system. Once again, the difficulty is caused by the multiplicity of units which make up a major industrial plant — and therefore of control systems needed. The consequent data processing has been made

---

* Also reviewed, in greater detail, in the companion book *Maintenance Organizations and Systems.*

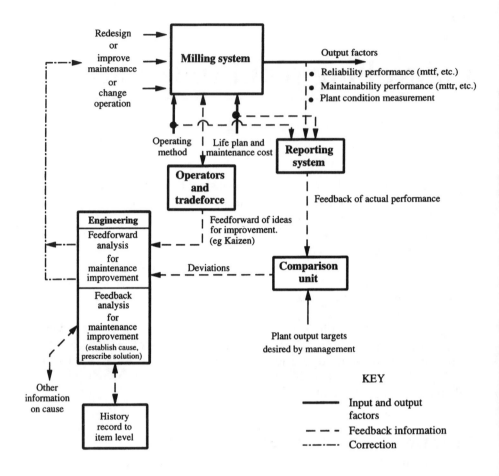

Figure 10.1. Controlling the reliability of a unit of plant

manageable by modern computer technology which can easily handle the many independent control mechanisms. The difficulty, however, usually lies not in the processing but in the acquisition of the data. Company management may therefore need to concentrate control effort on selected units, those which they deem critical; for the rest they may use the maintenance costing system to identify the most troublesome, e.g. those of highest high maintenance cost, poorest product quality, highest downtime, and so on.

Pro-active control of unit reliability

Figure 10.1 also illustrates the pro-active approach, which differs from the reactive in that it does not wait for failures or for high-cost problems to occur before taking action. The basic idea is that all members of the organization —

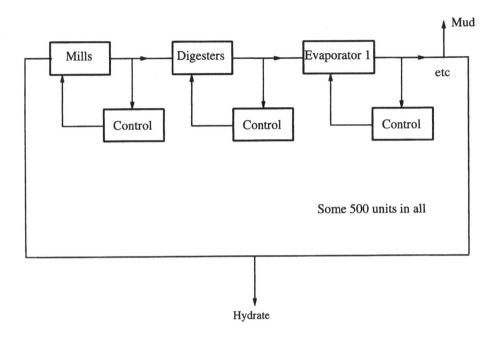

Figure 10.2. Controlling the reliability of a plant

but especially the shop floor — should continuously seek ways of improving unit reliability, and hence output, safety, and so forth.

The Japanese call this *Kaizen* (see Figure 14.2). The shop floor form small inter-disciplinary, but plant-orientated, teams to improve the reliability of selected units. (Preventive maintenance is interpreted literally — to prevent the need for *any* maintenance, by design-out and other actions.) Upper management circles ensure that the idea is promoted and accepted throughout the organization. This ensures that middle management circles give assistance and advice to the shop floor teams as necessary.

Incorporating reliability control systems into the organization

Although Figures 10.1 and 10.2 are useful for understanding the mechanisms of reliability control it still remains to incorporate these ideas into a scheme for a working maintenance organization. This is shown in the general model, Figure 10.3, and the application of this to the alumina refinery organization, which is outlined in Figure 10.4.

It can be seen that there are several inter-related levels of plant reliability control in an organization — each with its own responsibilities and roles. The first operates between the shop floor and supervisors and to a large extent is

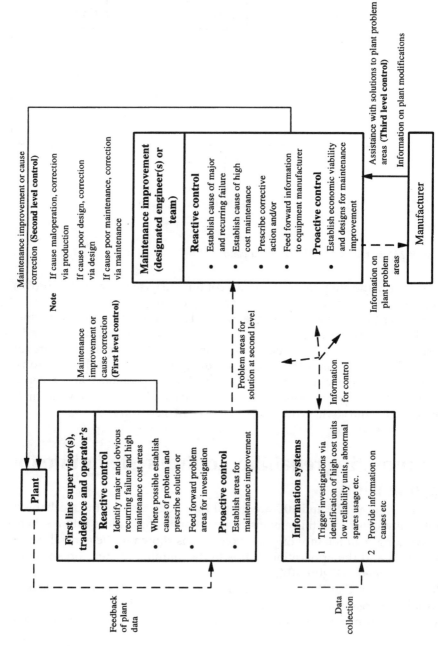

Figure 10.3. General model of reliability control within an organization

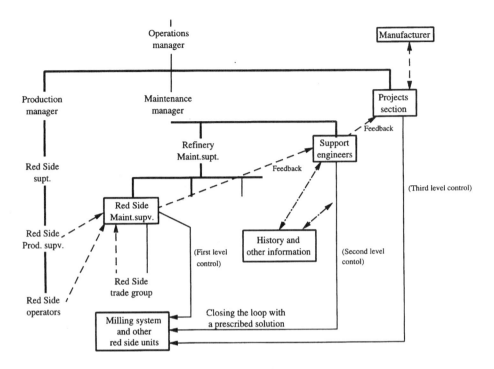

Figure 10.4. Plant reliability control within the refinery organization

independent of the information systems — however a history record can be important here. This level of control is particularly useful because there is a quicker reaction to problems. Because the personnel involved may be present at a repair, and can discuss it with operators and tradesmen, there is likely to be first-hand knowledge of the cause of failure. In addition it is at this level that the main thrust of pro-active control operates; if the personnel involved cannot establish the cause and/or prescribe and implement a solution, then the problem is passed up to the second level.

The second level of control operates through designated engineers and/or a maintenance investigation team. To be effective, this requires the integration of information systems and engineering investigation. The information system (computerized) should be designed around the ideas illustrated in Figure 10.5 and should therefore be capable of identifying problem units and hence triggering corrective investigation within the organization (satisfying Point (a) of the control requirements). The major effort, however, will lie in the diagnosis of failure causes (Point (b)), by interrogating the plant history, and in the prescription of corrective action (Point (c)), an effort which will need to come from the investigative engineers.

In general it is the root cause of any problem which will be sought, and because investigate effort is necessarily limited only a small number of problem items can be looked into at any one time. The criterion for selecting these

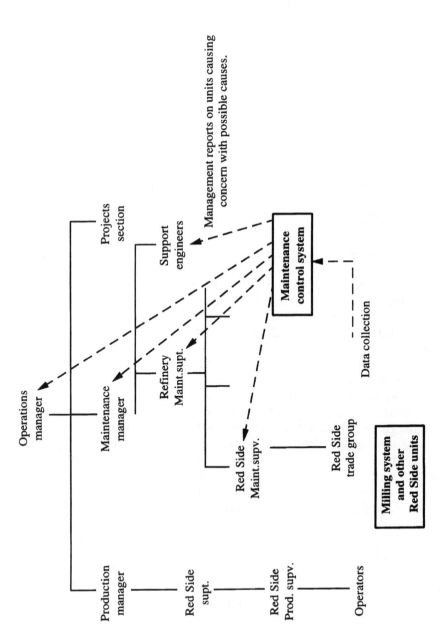

Figure 10.5. The use of maintenance information systems for reliability control

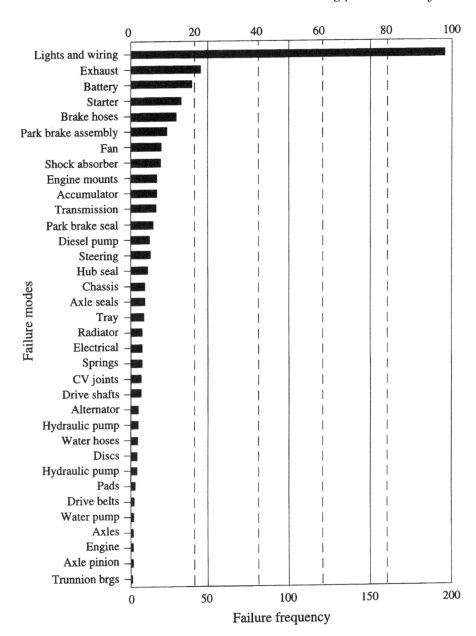

Figure 10.6(a). Pareto chart of failure frequencies, four wheel drive vehicles used in an underground mine

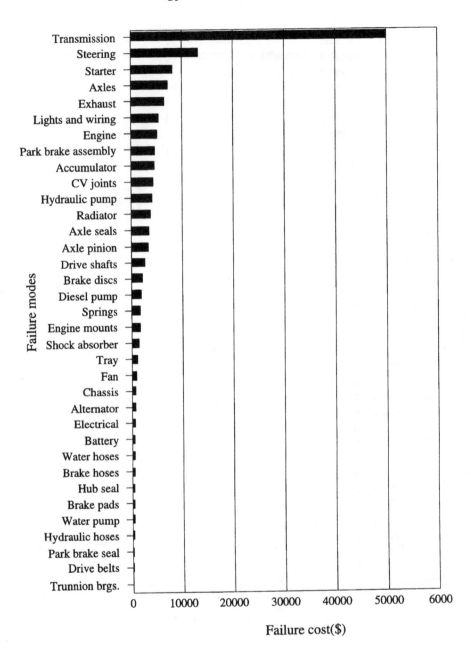

Figure 10.6(b). Pareto chart of failure costs, four wheel drive mine vehicles

is usually based on some kind of equipment ranking, by downtime, direct cost or failure frequency (see, for example, Figures 10.6(a) and (b) which show such rankings, Pareto analyses, for a mining vehicle fleet[1].

A possible third level of control lies in the contact between the various users of a given type of equipment and its manufacturer. This offers the opportunity for maintenance information to be collected from a much larger pool of experience. However, because more than one user company will, in practice, be involved it is the least effective level of control. The onus for ensuring the success of such an activity rests with the equipment manufacturer.

Reference

1.   Healy, A., *Effect of road roughness on the maintenance costs of four-wheel drives*, PhD thesis, Queensland University of Technology, Brisbane, Australia, 1996.

11
Case studies in maintenance strategy

Introduction

So far, the discussion has drawn on examples that have been taken mainly from various process industries. Further useful insights into the general problem of maintenance strategy formulation — its principles, concepts and modelling procedures — can be gained by looking at it in the context of other, quite different, technologies, such as mineral extraction, or electricity generation and supply.

Case study 1: A gas-fired power station

An illustration of the linkage between production policy and maintenance strategy. Because the first four case studies in this chapter are all derived from work in various electricity generation and supply systems it will be useful to begin by outlining the general operating characteristics of such systems.

Figure 11.1 is a schematic of a generation and supply system. Typically, demand for electricity varies throughout the year as shown in Figure 11.2. (The demand will also vary, of course, throughout the week — there will be less demand at the weekends — and throughout each 24 hours — there will be less demand at night.) Several generating units (GUs), of various sizes, will feed the distribution grid. The most efficient of these, usually the larger ones, will supply the base load (the non-varying demand), the less efficient ones being brought on intermittently to meet peaks in the demand. Not uncommonly, gas turbine and/or hydro units will be employed to meet demand peaks of *short* duration (a practice sometimes referred to as *peak lopping*).

The first case study is of a maintenance strategy review of a gas-fired power plant.

The station and its operating characteristics

The station concerned had an installed capacity of 600 MW(e), made up of five 120 MW(e) sets, each of which comprised a gas-fired boiler and steam-

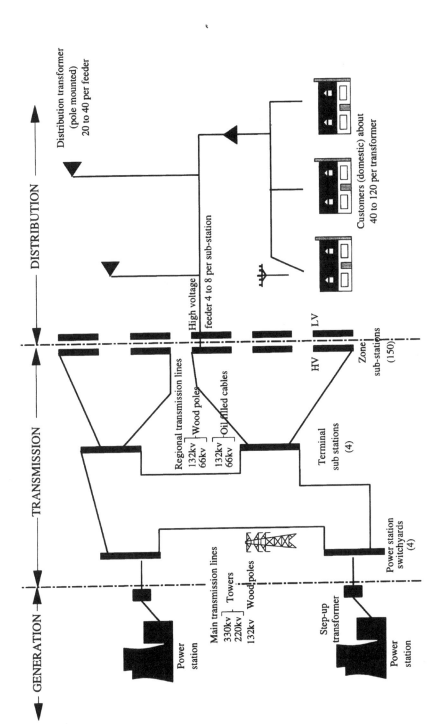

Figure 11.1. Schematic of electricity supply system

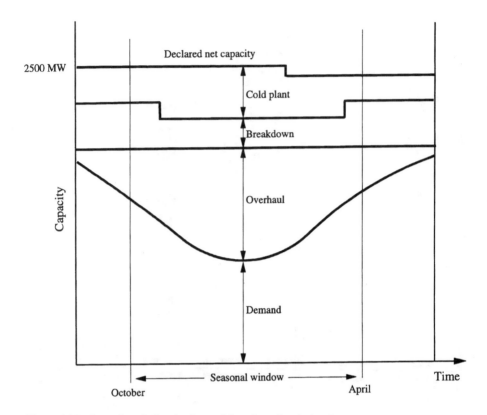

Figure 11.2. Annual variation in demand (southern hemisphere)

driven turbine. Until a year before the maintenance review the station had been part of the base load supply. It had then moved down in the merit table and at the time was being used on a two-shift operating pattern, i.e. it tended to be used each day from 6 am to 8 pm but was not required at night, when the demand fell.

Relatively little off-line work could be carried out at night because of shortage of time for cooling and isolation, and also because the station was expected to provide a 'spinning reserve'. However, production-related windows for one or more of the GUs occurred on a more random basis and could be up to two weeks in duration. Such windows occurred — mainly during the annual low demand period — on average, about three times per year per GU. The planning lead time for these randomly occurring windows was relatively short (about one week, at most).

The maintenance strategy in use when the station provided base load

The major-outage life plan for a GU when the station was operating to provide base load is shown in Figure 11.3. This programme was the main

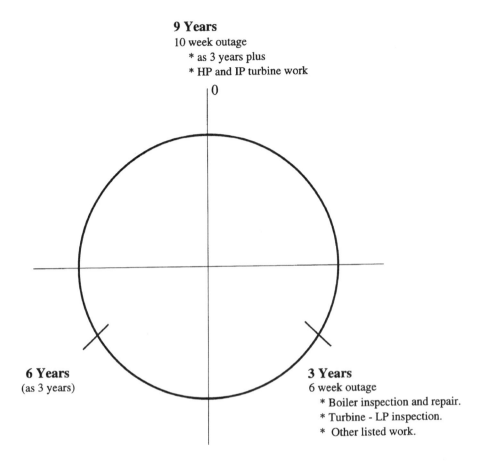

Figure 11.3. Outline of the major outage life plan for the generating units

thrust of a GU's life plan. Relatively little work other than lubrication and simple inspection was undertaken outside the major shutdowns. The *major shutdown schedule* took account of the pattern of grid demand and also of the availability of internal and contract labour. For the station as a whole there was a ten-year plan, a maximum of two units being overhauled in any one year. This generated a workload of the kind illustrated in Figure 11.4.

Maintenance strategy review for two-shift operation

The strategy was reviewed using the TDBU approach. The work content of a major shutdown was examined and, as far as possible, reduced by the following actions:

(i) jobs were identified which might possibly be scheduled into the randomly occurring windows;

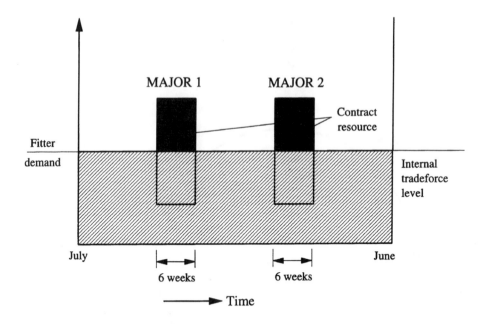

Figure 11.4. Workload pattern and its resourcing

   (ii)   jobs which could be carried out in windows provided by the presence of redundant or spare plant, e.g. work on duplicate pumps, were identified and rescheduled;

   (iii) on review, many jobs that were previously done at fixed intervals became condition-based ones; in some cases it was decided to do them only after failure.

The review resulted in an improvement of availability, mainly due to the reduction of the duration of scheduled outages (the maintenance transferred into the randomly occurring windows not then having any direct impact on availability) but also because the revised maintenance policy was more effective. In order to implement this revised strategy and, in particular, to facilitate the necessary opportunity-scheduling, improved work planning systems (based, among other things, on better computer software) were needed.

Case study 2: An oil-fired power station

This study will illustrate the linkage between production objectives and maintenance objectives and shows how pursuit of the latter can drive changes in the life plan and maintenance organization.

The station and its operating characteristics

An installed capacity of 360 MW(e) was achieved via five 60 MW(e) sets, using oil-fired boilers and steam-driven turbines, and a 60 MW(e) gas turbine. The steam-driven units were some thirty years old and the gas turbine 22 years old. By the time of the study the station was privately owned, having been run down — with a view to decommissioning — under its previous state ownership. The existing management had a contract to supply electricity using the steam-driven units until the year 2000 and the gas turbine unit until 2010. This would depend on many uncertain factors, among which were whether the local grid would be connected to other grids, the future demand for electricity, environmental legislation and so on.

The station provided a peak lopping service to the grid. For this, the gas turbine could provide an immediate response while the steam turbines could respond with as little as four hours' notice. The contract for the steam turbines was for four units out of the five — i.e. 240 MW(e) — to be available at any time. Thus, these units could be considered separately from the gas turbine as regards most aspects of maintenance strategy. The presence of the extra steam unit provided numerous windows for scheduling off-line maintenance work without losing system availability. Taking the gas turbine off-line for maintenance always meant, at any time, a total loss of *its* availability.

Production and maintenance objectives

Production objectives were determined by factors which had been set under contract. For the steam units, payment was based on availability rather than supply. Full payment resulted from achieving 100 per cent availability of four units (i.e. of 240 MW(e) capacity); various checks and penalties could then modify this. The availability actually achieved at the time of the study was about 98 per cent. For the gas turbine, payment was based partly on availability (e.g. 50 per cent of maximum payment could be obtained by achieving 100 per cent availability) and partly on operational reliability (e.g. 50 per cent of maximum payment could be obtained by achieving 100 per cent successful response to all the demanded starts). At the time of the study the gas turbine availability was over 80 per cent, its operational reliability of the order of 90 per cent.

Environmental and personnel safety standards were not discussed so it was assumed that they were satisfactory. A plant-condition audit was not carried out but it was known that the equipment was old and that, during the last ten years, it had been allowed to deteriorate. An important question was 'What was the expected remaining life of each steam unit, given its age and condition?' The answer to this would have a major influence on its maintenance life plan.

The management were aware of the above considerations and their inter-relationship. They had identified the maintenance objective as being:

to maintain or improve the present output performance of the generators while reducing the resource cost via improvements in maintenance organizational efficiency.

Maintenance strategy before privatization

Steam units
The life plans could be summarized as follows:

- A major outage of twelve weeks duration every six years, to carry out statutory inspections, boiler, turbine and ancillary equipment overhaul.
- A major outage of three weeks duration every 26 months. The frequency of the shutdown was that of the statutory inspection of the boiler but other necessary work was also carried out.
- An annual outage of 1½ weeks duration, to undertake boiler and turbine inspection and ancillary plant maintenance.
- On-line lubrication and simple inspection routines.

The station maintenance schedule was aimed at spreading the outages as evenly as possible over the six-year cycle, in order to smooth the station workload. Essentially this meant that, on average, there were 25 weeks of outage work per year. The station's internal maintenance labour was manned up to this shutdown workload. Little use was made of any contract resource.

Gas turbine
The life plan was built around a major outage, every four years, of six weeks duration and an annual outage of two weeks duration. Because of its specialized nature and the high cost of spares holding, this work was contracted out to specialists, except for the first-line work which was covered by internal labour. This policy remained the same after privatization.

Maintenance strategy after privatization

After privatization, considerable effort was devoted to changing the life plans and station outage schedule in order to maintain the steam unit availabilities and reliabilities at reduced maintenance cost. This was achieved via the following actions:

(i)   Discontinuing, after 1995, the six-yearly outage because the steam unit lives would come to an end by the year 2000. The remaining two twelve-week outages were scheduled for

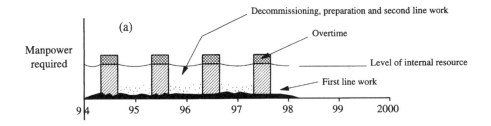

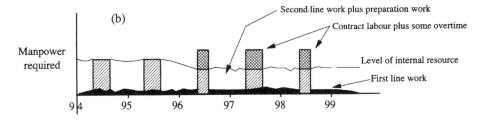

Figure 11.5. Workload and resourcing (a) before and (b) after strategy change

the summers of 1994 and 1995 (the last of these being a precautionary outage).

(ii)   Discontinuing the traditional annual outage and incorporating its work into the 26 monthly statutory outage. This extended the duration of that outage to four weeks. These outages were scheduled for the summer months at a rate of three in one year, two in the next year, and so on.

(iii)   Reviewing all the shutdown work to identify the jobs that could be undertaken outside the main shutdowns by taking advantage of plant redundancy. This work was incorporated in an ancillary equipment preventive maintenance programme (a 'window' schedule) and was scheduled to smooth the workload between outages.

The changes in workload and resourcing are shown in Figure 11.5. Peaks were resourced via a combination of contract labour (mostly) and overtime. The main benefit was a reduction of 40 per cent in the internal maintenance tradeforce and of an overall 30 per cent in labour costs. In other words, the change of strategy allowed an improvement in organizational efficiency without a loss of maintenance effectiveness.

The management of the station was also embarking on measures to reduce labour costs by improving flexibility, i.e. by reducing the non-trade workforce, improving inter-trade and operator-maintainer flexibility. This would lead to the same workload being carried out by less labour and at lower labour cost.

Case study 3: A transmission system

This will show how the TDBU approach (in particular Step 1) can be applied to non-process plant.

The transmission grid was outlined in Figure 11.1. Its function was to transmit power from the generating stations to the zone sub-stations and then to the local distribution systems. In order to transmit the power efficiently the station transformer stepped up the voltage to 330 kV; the power then going via the main switchyards to the grid.

The grid itself comprised main sub-stations, zone sub-stations, main transmission lines carried on steel towers, and regional transmission lines carried on wooden poles and in oil-filled underground cables (see Figure 11.1); these were the *primary assets*. In addition there were the following *secondary assets*:

> *Grid control system*, including the host computer at the control centre and the transducers, etc. at the power station and sub-stations. This was mostly solid state electronic equipment.
>
> *Communication systems*, including the grid protection communication system and the microwave systems which passed information from the transducers to the host computer.
>
> *Protection systems*, made up mainly of solid state electronic equipment that protected the power stations, sub-stations, etc.

The *maintenance objective* for the transmission grid could be expressed as for process plant, i.e.

> to achieve the agreed system operating requirements, with agreed and defined plant condition and safety requirements, at minimum resource cost.

The system operating requirements were set as 'transmission practice standards'. Supply reliability, and safety, were measured via various quantitative indicators, and targets were set based upon these indicators. These requirements could then be translated into user requirements at main asset (e.g. sub-station) level and used to develop maintenance life plans.

The top-down approach could be used to develop an *equipment criticality ranking*. For example, consider the outline, in Figure 11.6, of a part of the transmission grid. The thick line indicates a main transmission line that could be regarded as critical, in the sense that if it failed it would restrict the flow of electricity from the power stations. If this line were to be required for off-line work it would have to be taken off-line when one (or perhaps two) of the GUs were on outage maintenance. This particular transmission

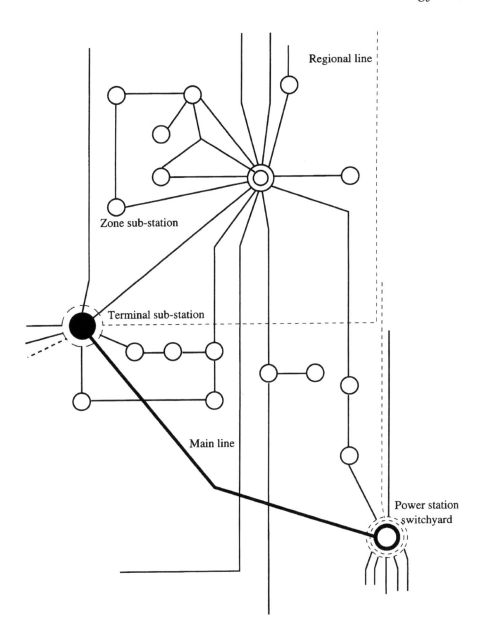

Figure 11.6. Part of the transmission grid

system was audited and one of the main resulting observations was that more attention would have to be given to identifying and ranking those lines, switchgear and failure modes which were critical to system reliability or safety. Such a criticality ranking was probably understood but, as far as could be seen, had not been documented as a part of the transmission system maintenance strategy.

The maintenance life plan for the *primary* assets had been developed in the conventional way outlined in Step 2 of the TDBU approach. The plan for a typical sub-station, for example, was as outlined below.

Inspection and lubrication routines	—	Monthly
A-grade service (a combination of inspection, proof testing, minor adjustment, replacement of simple items)	—	Three-yearly
B-grade service (broadly similar to A-grade)	—	Six-yearly
Overhaul	—	Based on the results of the services

In general, the life plans for the *secondary* assets were different because they were largely solid state electronic equipment . The plans were therefore based on routine cleaning and calibration, some proof testing and some planned corrective maintenance.

The maintenance schedule for the main lines was driven by the outage requirements for the GUs. This in turn drove the outage schedule for the switchyards, main terminals and sub-stations and hence influenced the schedule for the secondary assets.

Although this scheduling seems straightforward it should be appreciated that the GUs, main lines, regional lines, switchgear and secondary assets were the various responsibilities of different parts of a large organization. Thus, the effective co-ordination of effort throughout such a large organization required excellent communication systems — which is a story for a later chapter.

Case study 4: A distribution system

This study will complete discussion of the generation and supply system of Figure 11.1. It is instructive because it will show that, even if an otherwise satisfactory maintenance strategy is being followed, preventive maintenance can be neglected if objectives and work priorities are not clearly laid down beforehand.

The distribution system boundary was at the zone sub-station of Figure 11.1. Even in this example's relatively small grid of 2500 MW the size of the

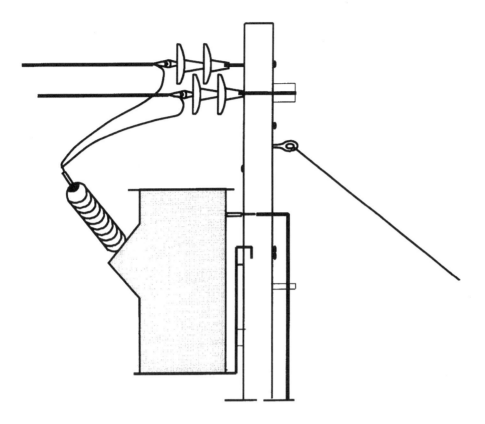

Figure 11.7. Pole-mounted transformer

distribution assets was considerable, namely 150 sub-stations and thousands of wooden poles each one carrying some appropriate equipment (such as the distribution transformer shown in Figure 11.7).

Although both corporate and maintenance objectives had been specified for the generation and supply systems they had not been interpreted into objectives for the distribution system. Simple life plans — broadly similar to those for the transmission sub-stations (i.e. based on inspection-oriented services) — had been formulated for each of the sub-stations. In general, there were life plans for the pole-mounted equipment, based on simple inspection when the wooden pole structure was being maintained (which, because of the age of the poles — as much as thirty years — and the prevalence of adverse ground conditions, was itself the main source of work). The maintenance policy for the pole structure was as indicated in Figure 11.8. An additional maintenance task in some areas was tree clearing around the lines.

Audit of the distribution system maintenance revealed a backlog of work on the poles, and on the equipment mounted on them, which was

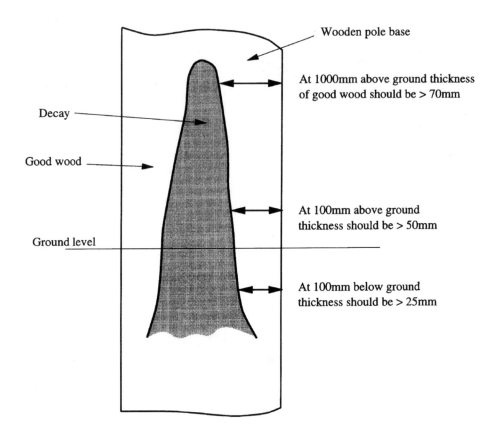

Wooden pole base

At 1000mm above ground thickness of good wood should be > 70mm

Decay

Good wood

At 100mm above ground thickness should be > 50mm

Ground level

At 100mm below ground thickness should be > 25mm

Four/yearly inspection, based on the following

* If below-ground criterion is met then chemically treat base only.

* If below-ground criterion is not met but upper criteria are met then reinforce the base with steel stakes and chemically treat.

* If lower and upper criteria are not met then replace pole.

Figure 11.8. Pole maintenance policy

many years long. The condition of these assets was clearly deteriorating and causing senior management concern, both for safety and for security of supply. The basic cause of this problem can be deduced from Figure 11.9, which shows the maintenance workload for a typical distribution area. The tradeforce had to carry out not only the maintenance of the existing network but also the expansion of the network to new homes and industries

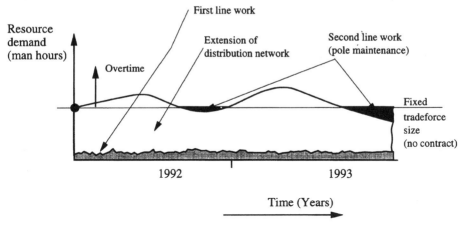

Figure 11.9. Long-term workload for line tradeforce

— which generated new income and therefore took priority. The absence of asset condition and safety standards led to continued deferral of maintenance. At best it was carried out only when expansion work eased off.

The following recommendations resulted from the audit:

- Distribution maintenance objectives should be set and translated down to main asset level. Safety and longevity standards should be specified.
- Either a separate 'maintenance group' should be formed or work priorities changed so as to ensure that maintenance is carried out at the required time.

Case study 5: A petroleum refinery

The author has audited the maintenance of several petroleum refineries. This study of strategy formulation highlights maintenance problems that are typical of such plant.

The plant and its operating characteristics

A simplified process flow diagram for the refinery is shown in Figure 11.10. The plant had been on the same site for over forty years and in that time had had major extensions and modifications. Each of the plant sections shown could be represented by a process flow diagram analysed to unit level, as in Figure 11.11.

Scheduling characteristics

The plant was production limited. Thus, for the foreseeable future there were

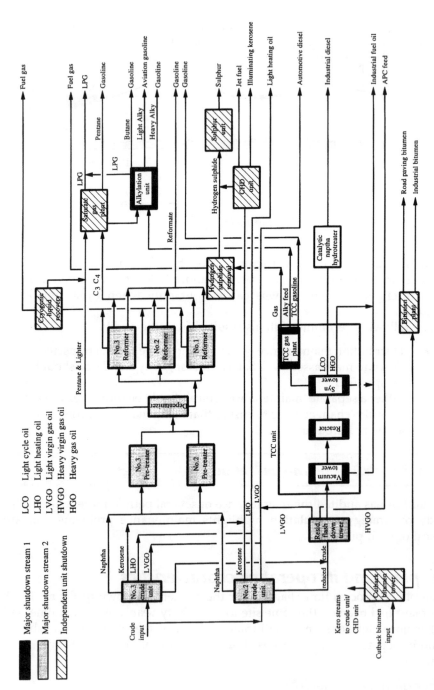

Figure 11.10. Refinery process flow

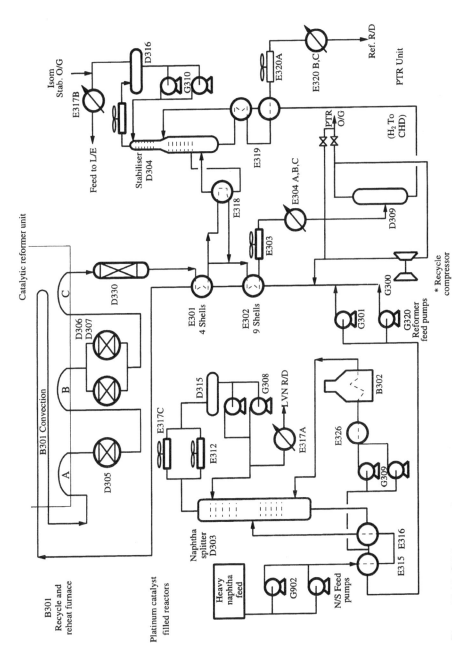

Figure 11.11. CFU process flow

no plant-level production windows. The plant consisted of two main process streams one of which included the crude unit (CU) and the other the thermal catalytic cracker (TCC). Several other plant sections, e.g. the sulphur unit, could be taken off line while the rest of the refinery continued to operate. With the process streams it was not possible to operate (or operate effectively) with one of their major plant units down. However, there was a considerable level of redundancy at item level (e.g. many pumps were paired, one being normally on line, the other on standby). There were also several windows at unit level caused by production maintenance, e.g. by catalyst changes.

Critical plant

Could be considered as any item where failure would affect product output or quality (e.g. the recycle compressor of Figure 11.11) or would create an immediate or potential safety hazard. Since this was a plant that handled hazardous chemicals there were many safety critical items (e.g. all the safety relief valves and any pump or vessel from which there could be leakage of flammable or toxic fluid).

Maintenance strategy

This could be summarized as follows:

A schedule of outage work for the main process streams

The CU stream was shut down for a 28-day overhaul — based on the statutory pressure vessel inspections — every four years, and the TCC stream for a 28-day overhaul every two years (the estimated time for various wear-out effects to become significant). The remaining units came out independently at intervals appropriate to their optimum running times. The work content of these outages comprised the following actions:

- Condition-based maintenance prompted by information from previous shutdowns or from on-line monitoring.
- Condition-based maintenance prompted by inspections carried out in the current shutdown.
- Fixed-time repairs and overhauls.
- Deferred corrective jobs.

In practice, more than 30 per cent of the outage work originated from current outage inspections. This was because information from on-line inspections (which were themselves less than satisfactory) and from previous shutdowns was poor.

A schedule of maintenance work for the stand-by equipment (e.g. for pumps)

In general, such a schedule did not exist. The pumps were either operated to failure or replaced via *ad hoc* operator monitoring. The operating times of the pumps were not recorded and there was no set operating policy for pump systems (e.g. 'Operate one pump to failure and keep the stand-by pump as new').

On-line-inspection routines
Dynamic plant. Several key units (e.g. various compressors) had a fixed vibration monitoring system in place. Any other such monitoring was requested on an irregular, *ad hoc*, basis.

Stationary plant. A non-destructive-testing key-point inspection programme was computer scheduled.

Simple inspection routines. Although the operators were the equipment owners most of their inspection was process-oriented. Little effort was being devoted to simple maintenance inspections.

Observations

The main thrust of maintenance strategy for large process plant lies in the major outage (or turnaround) work. The maintenance policies, work content and frequency of this have usually evolved over a long period of time — often leading to the establishment of recognized standards and codes of practice — and are usually satisfactory. This plant was no exception and, as a consequence, the availability performance was good by international standards. With better on-line monitoring, however, and better information from previous shut-downs, it should have been possible to pre-plan 90 per cent of the outage work. This would have resulted in better shutdown planning, improved work quality, more appropriate spares provisioning and therefore shorter shutdown durations.

The non-outage preventive maintenance schedules (mainly of routines for ancillary equipment such as standby pumps) were poor. The main reason was that the life plans for such equipment had not been systematically established and documented. (The author has found this to be the case with many of the refineries and other large process plants that he has audited.)

All of the main ancillary items of plant should have had a properly reasoned operating policy and maintenance life plan. Such a document should include:

(a) For standby items, a policy specifying operational checks and the recording and monitoring of running time.
(b) A lubrication routine and operator-inspection checks.
(c) A service schedule, specifying — where necessary — vibrational or other condition monitoring techniques.
(d) A reasoned repair/replacement procedure based on recorded running time as specified in (a), or on observed condition as specified in (b) or (c).

Case study 6: An open cast mine

This will show how even a mine can be modelled as a process flow system. It will also provide an introduction to the concepts and principles of fleet maintenance.

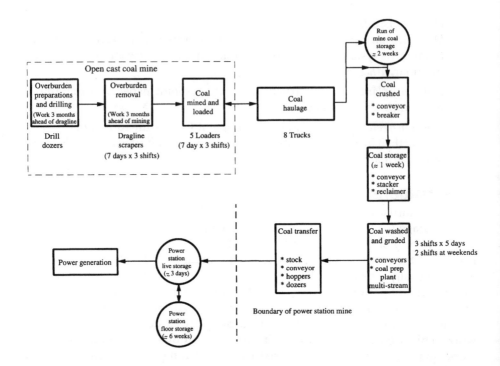

Figure 11.12. Process flow of open cast coal mine

The plant and its operating characteristics

The particular open cast mining installation is outlined in Figure 11.12. The process started with the stripping and removal of the overburden (the soil and rock above the coal seam) using drilling, explosives and a dragline. The exposed coal seam (metres thick) was then extracted and loaded into trucks for haulage to the coal preparation plant. The coal was crushed in the first operation and washed and graded in the second one before it was finally conveyed to the rail head for transportation. There were a number of points of inter-stage storage and also final product storage. This gave operational flexibility to each individual process and also to the operation as a whole.

The main feature was that the process depended to a large extent on the performance of small fleets of diesel powered equipment. For example, four front-end loaders for the mining operations and eight large dump trucks for the haulage operation. (Other mobile equipment included dozers, scrapers, graders, drills, etc.) It is this equipment — rather than the fixed or semi-fixed units such as coal washing plant, draglines, etc. — that caused the maintenance problems and was the subject of this study.

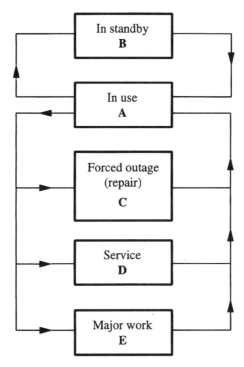

Haulers - 21 shift operation. Operations require minimum of 5 out of 8
Loaders - 21 shift operation. Operations require minimum of 3 out of 5

Figure 11.13. Status diagram for mobile fleet

Modelling fleet operation and maintenance

Each of the fleet types (e.g. the loaders) can be represented by a status diagram as in Figure 11.13 (i.e. a loader can be in any one of the states A to E). The availability of a single item can be measured conventionally, i.e.

$$\text{Availability} = \frac{\text{Time up}}{\text{Time up} + \text{Time down}}$$

$$= \frac{\text{Time in A} + \text{Time in B}}{\text{Time in A} + \text{Time in B} + \text{Times in C, D and E}}$$

Such a measure is useful since it provides an index of effectiveness of the maintenance effort for that unit. It also provides a comparison with the manufacturer's specified availability, which is usually defined in a similar way. For the loaders, the manufacturer had quoted an availability of 85 per cent if the equipment was operated correctly and maintained according to his recommended life plan.

However, the important index for the small fleet of loaders was the proportion of the fleet, the fleet demand ratio (FDR), that was required by Production to be in operation at all times during production shifts. A minimum of three loaders had been specified, and to satisfy this — and to carry out maintenance in states C, D and E — the company carried a fleet of five. Thus, the FDR was three out of five, or 60 per cent.

It appeared, from the manufacturer's availability figure, that the company had played safe. However, the decision to carry five loaders rather than four was based on the following influencing factors:

- The operation was production limited and a high downtime cost would be incurred if the number of loaders operational were to fall to two.
- Production wanted cover when a loader would be undergoing major overhaul (every two years) or major repair after failure.

The specification and measurement of availability ratios could have been usefully supplemented by some monitoring of the level of in-service failures. The best way of doing this would have been to keep a simple count as a function of shift, day, unit number, unit type, etc. (It could be argued that unavailability costs did not occur in the same way as with fixed plants; they had been 'bought off' in the capital cost of the extra fleet capacity.)

Observations

An interesting point is that the maintenance supervisors felt that their objective should be 'to ensure Production a minimum of three loaders at all times, at minimum maintenance cost'. The maintenance manager, however, felt that it should be 'to achieve a loader minimum availability of 85 per cent, at minimum maintenance cost'. His view, which the author sympathized with, was that such an availability would also meet the production requirement. The availability or reliability of mobile mining equipment is normally less than that predicted by its manufacturer, and its maintenance costs often considerably higher. There are several common reasons for this, namely

(a)   The equipment selected may not have been the most appropriate for its duty.

(b)   For many reasons (e.g. pooled use) there is little sense of ownership by its operators. This leads to much maloperation, especially when coupled, as it often is, with severe operating conditions and bonus payment arrangements.

(c)   For several reasons — including (a), (b) and poor maintenance organization — the preventive programme is often neglected, equipment condition deteriorates and more corrective work is needed. This in turn results in even less preventive work being done — and so on until the whole fleet maintenance becomes purely reactive.

All of this was evident in this particular case. Clearly, the long-term solution was to upgrade the fleet with new, wholly appropriate, equipment. Recommendations for the shorter term included:

- Improving the sense of ownership by allotting operators — who would carry out simple pre-shift inspections and other minor maintenance — to equipment. Improving their training in both operation and maintenance, this to include improving their understanding of the links between symptoms and the failures.
- Carrying out a condition audit of the existing equipment and establishing a corrective maintenance programme to bring the equipment up to an acceptable condition (a 'catch up' strategy). In conjunction with this corrective effort, reviewing and modifying, as necessary, the equipment life plans.
- Changing the maintenance organization to set up a preventive and overhaul group and a corrective group. Ensuring that the work planning system and its priorities reflected the importance of the preventive programme.

Case study 7: A local passenger transport fleet

This last case study will reinforce the principles and concepts — as introduced in the previous study — of fleet maintenance. The transport authority concerned operated from twenty or so garages located in different parts of a large conurbation, each providing transport in its own area and also the necessary parking and maintenance facilities. The garages were divided into three groups and in each one the major maintenance work (overhauls) and reconditioning was carried out at a central works. The system for a single garage and works is outlined in Figure 11.14.

The buses employed were mostly of the double-decked, front entrance, rear-engine type. The various models are enumerated in Table 11.1, which also shows the peak demand. The existence of a surplus of vehicles — i.e. above this peak level — provided a small standby pool on which essential maintenance could be carried out.

At the time of the study, the maintenance life plan for a bus was based on inspection and service at three-weekly intervals (3000 miles). Shorter and long-term maintenance schedules were based on sub-multiples or multiples of this basic period. At the end of the year the bus was prepared for its annual MOT test. Overhauls were carried out at intervals of approximately three years. This plan had evolved over a period of time and was in need of review because it was felt that:

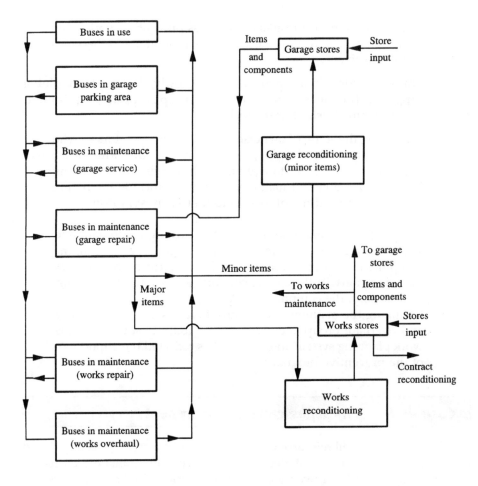

Figure 11.14. Status diagram for a large passenger fleet and its maintenance system

Table 11.1. Fleet inventory and maximum demand

Bus type	Bus make	Number in fleet	Maximum demand
Single deck	A	6	4
Double deck	B1	1	
	B2	14	
	B3	16	
	B4	13	
	B5	40	
	C1	23	
	C2	86	
	D	10	
		203	177
	Total all buses	209	181

(i)   the fleet demand ratio (FDR, see previous case study) was too low;

(ii)   the incidence of in-service failures (and unscheduled corrective work) was too high;

(iii) the existing inspection procedures were too subjective and often not carried out.

The TDBU approach was used to review the strategy for the fleet maintenance operation. This discussion will be confined to those sections of the review which were special to fleet operations.

Step 1: Top-down analysis

In the analysis of the previous case study it was assumed that the production demand for fleet units was constant, i.e. it was always for a minimum of three. In most fleet operations however, it fluctuates with time and the bus fleet of this case was no exception (see Figure 11.15, which indicates numerous 'production windows'). Because of the difficulty and expense of night-time and weekend working the most convenient windows occurred midweek — from 9 am to 4 pm — and facilitated the routine inspections and servicing, and other minor maintenance.

Although time for maintenance was available *outside* these midweek windows, the number of buses being 16 per cent in excess of the peak demand,

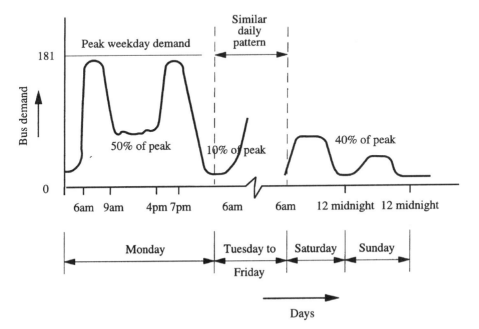

Figure 11.15. Bus demand pattern

it was clearly advantageous to try to make maximum use of the opportunity they provided because this would reduce the need for excess buses and hence the capital cost of the fleet.

Step 2: Bottom-up analysis

The existing life plans for each bus type — which had evolved via custom, practice and manufacturers' recommendations (and were felt, by the supervisors and tradeforce, to involve a degree of over-maintenance, especially as regards routine servicing) — were reviewed using an analysis of the kind outlined in Chapters 8 and 9. The main thrust of this exercise was to extend the basic service period from three to six weeks (see Tables 11.2(a) and 11.2(b)) and to move towards thorough and comprehensive inspection procedures (see, for example, the basic six-weekly checks listed in Tables 11.3(a) and 11.3(b)).

Table 11.2. Revised bus life plans: (a) minor work; and (b) major work

Table 11.2 (a). Minor work

Work and frequency	Outline description
Daily checks	Tyre pressure, engine oil, cleaning, etc.
Weekly safety checks	Steering gear and lubrication, etc.
Six-weekly service	Basic service, see Table 11.3
Twelve-weekly service	Basic, plus engine oil change and oil analysis check. Also valve clearances and fuel cylinders
Eighteen-weekly service	Basic, plus gearbox oil change and oil analysis. Gearbox calibration and bearing adjustment on front axle
Twenty-four-weekly service (Continued to 48 weeks in multiples of six weeks, then repeated)	Twelve-weekly service plus checking of fuel system, king pin bushes, gearbox, piston seals

Table 11.2 (b). Major work

Frequency	Inspection and corrective maintenance	Inspection time	Duration (weeks)
Three-yearly	Complete bus inspection and repairs at works. Thorough inspection/replacement/repair of all items and/or components. On completion bus to undergo a Freedom from Defects test	Four to five hours	Ten
Six, nine and twelve-yearly	Same as three-yearly	Same	Ten
Fifteen-yearly (economic life of bus)	Fixed by management on a criterion based on a combination of economic obsolescence, and condition factors		

Table 11.3. Revised basic six-weekly servicing:

(a) Inspection

Check all engine mounts and tighten
Check and torque cylinder head nuts
Check operation of fuel injection pump
Drain and refill cam box
Check lubrication system for oil leaks
Check and adjust radiator fan belt tension
Check exhaust manifolds for leaks and torque nuts to 25 lbf.ft
Check front axle ball joints for free play
Check and tighten clutch fluid coupling drain plug
Check engine–clutch fluid coupling–gearbox alignment
Check gearbox–angle drive alignment
Check propeller shaft joints and splines for wear
Check propeller shaft needle roller bearing for wear-lift up
Check propeller shaft and check amount of free play
Check propeller shaft circumferential movement
Check propeller shaft for noise and vibration
Check electro-pneumatic unit for leaks using shock pulse meter (SPM)
Drain away water from electro-pneumatic unit (drain plug)
Check brake and transmission system and items for air leaks using SPM

Table 11.3. Revised basic six-weekly servicing: (a) Inspection (*cont'd*)

Check brake liners for wear and damage
Check brake liner–drum clearance using feeler gauge
Check and lubricate automatic slack adjusters
Check starter motor commutator and brushes
Operational test of all electrical items in alarm and warning system, lighting
    system, trafficator system, start and stop system, and auxiliary items
Check security and rubber buffer of suspension springs
Check shock absorber fluid level, link rubbers, leaks
Check torsion bar stabilizer for security, damage and distortion
Check body panels for damage and loose riveting
Check chassis frame for damage and security of attachment
Check to ensure all autolube chassis lubrication bearing points are not clogged
Check alcohol evaporation unit strainer (winter only)

Table 11.3 (b). Lubrication

Grease water pump bearing
Grease fan shaft joint and splines, and fan centre bearing
Grease power steering ram ends
Grease propeller shaft splines and joints
Lubricate footbrake pedal linkage
Top up shock absorber level
Check, clean or replace heater and demister filters

Step 3: Scheduling the work

Servicing and minor repairs

Carried out in the respective garages, the daily, weekly and fortnightly services were simple checks which involved little resultant corrective maintenance and therefore required little time. Such work could be carried out by a separate service team in the 9 am to 4 pm window, or at night. The six-weekly maintenance was scheduled by dividing the year into eight six-weekly periods (five working days per week), leaving two weeks for the holidays and two weeks for MOT preparation. Because the total number of buses was 209, this required seven buses to be serviced per day. The estimated time for each service is shown in Table 11.4, and the daily loading — which does not take into consideration the resulting corrective work — in Table 11.5. In the majority of cases the servicing, and any corrective maintenance, could be completed within the window; where this would not be possible the bus would not be available to meet the peak demand and this would count against the peak demand ratio (PDR). In addition, this planned workload could be augmented by the unscheduled corrective work resulting from in-service failure, which could be minor or could demand several days' effort.

Table 11.4. Maintenance service type

Service period (weeks)	6	12	18	24	30	36	42	48
Estimated time (h)	$\frac{1}{2}$	$1\frac{1}{2}$	$1\frac{1}{2}$	3	$\frac{1}{2}$	$1\frac{1}{2}$	$\frac{1}{2}$	4
Maintenance class	A	B	B	C	A	B	A	D

Table 11.5. Daily workload for red group

| Maintenance period (weeks) | Maintenance class | Red group — 35 buses | | | | |
		Day 1	Day 2	Day 3	Day 4	Day 5
6	A	✓	✓	✓	✓	✓
12	B	✓	✓	✓	✓	✓
18	B	✓	✓	✓		✓
24	C	✓	✓		✓	✓
30	A		✓	✓	✓	✓
36	B	✓		✓	✓	✓
42	A	✓	✓	✓	✓	
48	D	✓	✓	✓	✓	✓
Buses per day		7	7	7	7	7
Hours per day		12.5	11.5	10	11.5	12.5

Overhauls and major repairs

These were undertaken in the central maintenance workshops. As shown in Table 11.2(b), the timing of major work was governed by the 'freedom from defects' (FFD) test and by the three-yearly overhaul. Thus the buses could be scheduled for overhaul and FFD test, at the central workshops, on a three-yearly basis. Taking into consideration the time (ten weeks) needed to carry out an overhaul, about fifteen buses would be in the works for overhaul at any one time. As before, this would count against the PDR.

Observations

Would the new plan reduce the combined costs of unavailability and of resources used? The daily, weekly and six-weekly scheduled preventive work would be carried out in the windows and would therefore not affect the PDR — and would not involve an increase in the workload. Changes in the major preventive work would be small and they also would have little effect, therefore, on the PDR or on the resources used. The most important point was whether the revised inspection and servicing procedures would lead to fewer in-service failures and less unscheduled corrective work. The level of such work that would result from the new plan was difficult to estimate. The more thorough consideration of the maintenance procedures for each item, the resulting increase in the number of items covered, and the greater objectivity of inspection procedures should result in a reduction of corrective work. Even a small reduction would result in fewer buses being in repair, a higher PDR and a smaller workload.

Three additional recommendations were made.

- The three-yearly overhaul period should be extended, initially to four years and, after experience with this period, to five.
- The time required for an overhaul (ten weeks, sometimes more) was excessive. The procedures should reviewed with a view to its progressive reduction.
- The economic life of the buses should be extended to at least twenty years.

These last were not accepted. At the time, the bus company was publicly owned and controlled. It has since been privatized and these and many other changes (not all positive) have been implemented.

12
Exercises in maintenance strategy

Introduction
This section of the book is rounded off by giving the reader the opportunity to attempt three exercises, each of which is aimed at testing his understanding of some of the strategic ideas that have been presented. Solutions to the exercises are given at the end of the chapter.

Exercise 1.   An aluminium smelter: the anode-making plant

Background
Figure 12.1 shows the process flow of a plant which produces carbon anodes for an aluminium smelter. The plant can be considered as being made up of three separate, different, but inter-related sections or processes linked to each other by inter-stage storage. This exercise concerns the front-end process, manufacture of the unbaked or 'green' anode blocks (see Figure 12.2), which are subsequently baked in the ring furnace (the second process) and then rodded (the third process). After this the finished rodded anodes go into store (which can hold about 24 hours' worth of smelter demand for anodes) and from there to the smelter pot rooms.

Between blockmaking and baking, there is a capacity for storing green anodes of approximately two weeks' worth of demand. In addition, the rate of production in the block making plant exceeds the rate of baking — so the blockmaking section operates continuously for fifteen shifts, Monday to Friday.

The smelting operation* is continuous and there are no smelter-level windows of opportunity for maintenance. The blockmaking section, however, does present such windows at weekends (which can be extended by exploiting the inter-stage storage and the ability to raise the blockmaking production rate). The production department have indicated that they want a high level of reliability over the fifteen-shift operating period (because failures affect the product quality).

---

* An electrolytic process for reducing alumina to aluminium. Carried out in cells called 'pots', hence the 'pot room'.

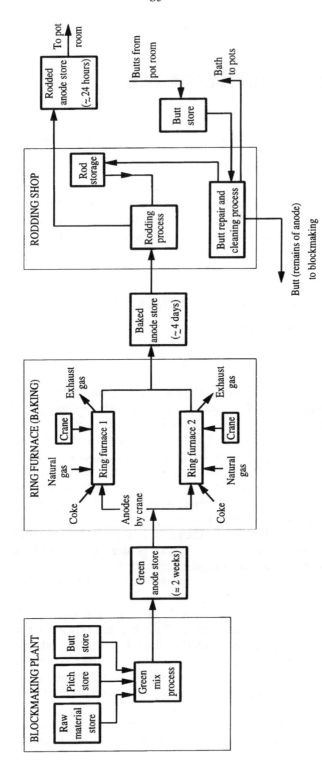

Figure 12.1. Process flow, carbon anode plant

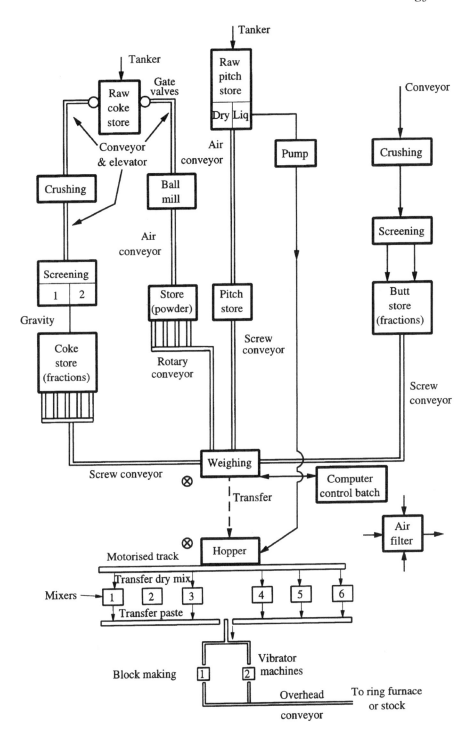

Figure 12.2. Process flow, blockmaking plant

A great deal of data has been accumulated on wear rates, failure incidence, etc. Most of the units shown in Figure 12.1 exhibit failure modes — the result of wear, corrosion, fouling, etc. — which are time-dependent.

In general, the unit life plans are based on a condition-based-maintenance policy which can be summarized as follows:

- Extensive on-line inspection routines involving

  - checking by operators,
  - tradesmen line patrol,
  - sophisticated monitoring techniques.

- A schedule of off-line inspection, at unit level. This work is aimed mainly at those units of plant deemed critical because of their significant potential, should they fail, for interrupting production or threatening safety. In general, the inspection lead time allows the repair or replacement work to be scheduled at least a week ahead. The largest off-line job resulting from inspection could be carried out by a team of about eight tradesmen in three or four days.
- For all other units of plant 'minimum maintenance' is carried out. In most cases this involves only lubrication routines.

The problem

Although there is no stated maintenance objective it is understood that 'Production want a high level of reliability over the fifteen-shift operating period and that Maintenance are expected to do this at least cost'. The present situation is that Production consider the availability and reliability levels satisfactory. Senior management, however, are unhappy with the high direct cost of maintenance in this part of the plant.

The recently appointed Smelter Engineering Manager considers that the existing maintenance strategy is not the 'best'. His opinion is that the life plans for the units should be based on off-line major maintenance carried out at fixed intervals (operating hours or calendar time) determined in the light of the extensive and detailed plant history. This would also allow the off-line work to be scheduled throughout the year to smooth the workload.

The questions that arise are the following:

(a) How does the existing condition-based strategy affect the maintenance workload? Does this present organizational difficulties?
(b) How do you think the proposed fixed-time approach will affect the reliability and availability of the blockmaking plant?
(c) How do you think the proposed fixed-time approach will affect the direct maintenance costs?

Exercises in maintenance strategy 209

(d) In view of your answers to (a), (b) and (c) how would you advise the Engineering Manager?

(e) The Engineering Manager feels that the plant control systems — including the solid state electronics, e.g. the PLCs — should also be put on the fixed-time maintenance schedule. Do you think he is right? If not, explain what you would do.

Exercise 2. A gold mine milling process

Background

Figure 12.3 shows the process flow of the milling plant of a gold mine. The mine is decoupled from the milling plant by the inter-stage ore storage. The milling process is the mine's rate-determining process. For the foreseeable future (the next five years) Management want to operate the milling process continuously, so there are no plant-level windows of maintenance opportunity. Downstream from the cyclone towers the off-line maintenance can be scheduled at unit level by exploiting redundancies (e.g. at any one time only three out of the available five thickeners are required).

Scheduled off-line maintenance, or failure of the crusher circuit, can stop the whole plant, although the plant can then be kept going for three days via the 'alternative' crushing process, but at four times the cost of normal crushing. Scheduled off-line maintenance or failure of a ball mill (or its ancillary equipment) causes a 50 per cent loss of milling production. Most of the maintenance results from time-dependent mechanisms, e.g. wear, corrosion or fouling.

The crushing circuit has a mean running time to failure of six months but failure predictability is poor because many of the failures are induced by randomly occurring production events. The ball mills also have a mean running time to failure of about six months but with good failure predictability. The main items needing replacement are the rubber lifters and liners.

The location of the mine is such that contract labour is extremely expensive. The resident labour force is manned up to the peak off-line maintenance workload and hence is not very productive for most of the time. The quality of the labour is good, with an excellent knowledge of the plant — and in particular of the corrective maintenance methods. Because of the high cost of production downtime the maintenance objective appears to be to maximize milling plant availability.

The existing maintenance strategy is based on the following actions:

• The crusher circuit is operated to failure (or near-failure as indicated by the operators' informal monitoring). Since failure

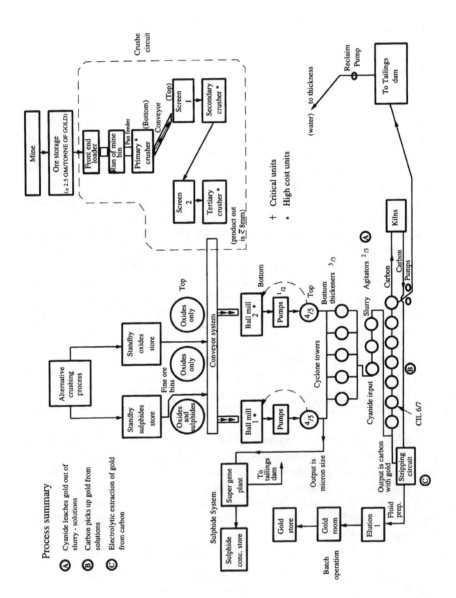

Figure 12.3. Process flow, gold mine milling plant

is expected there is a considerable level of pre-planning (e.g. preparation of spares, job methods, decision guidelines). Inspection and opportunity maintenance is also carried out on the other units of the crusher circuit not involved in any particular failure, and plant operation is sustained via the alternative crushing process.

- The ball mills are on a schedule of four-monthly overhaul. The main job is the repair or replacement of the lifters and liners, but other work is carried out on the mill to ensure its reliable operation over the following four months. In addition, preventive maintenance is carried out on other units in the stream, e.g. the conveyers. This causes a workload peak and contract labour has then to be employed. Some of the work is time based, some deferred, but most is repair-on-inspection.
- A minimum level of maintenance (mainly lubrication) is carried out on all other units.

NB Apart from informal monitoring by operators and by the Maintenance Supervisor little inspection work is undertaken.

The questions

(a) What would be the difficulty in using a fixed-time-maintenance approach for the crusher circuit?

(b) Do you consider that there is a better way of maintaining the crusher circuit? Outline your approach?

(c) While the existing fixed-time approach (four-monthly shutdowns) for the ball mills may not be the *best* procedure it is regarded as an *effective* one. Explain why this is so. How do you think this approach could be improved?

Exercise 3. An alumina refinery

Background

This example was used in Chapter 7 as a vehicle for discussing maintenance objectives. Some of the process descriptions will be repeated here for convenience.

The operating characteristics of the refinery can be derived from Figures 7.1 and 7.2. Figure 7.1 shows the relationships between the mine, power station, refinery and the various transportation systems and indicates that, as regards alumina production, the refinery is the rate-determining process. The refinery and the power station are integrated from a production point of view and both are therefore production-critical.

Figure 7.2 models the refinery process flows at plant unit level (a bauxite mill, for example, is represented as a single unit). Units incurring high maintenance cost or exhibiting low reliability are indicated. The refinery can be thought of as a series of process functions operating on a circulating working fluid. Bauxite and caustic are added at its front end and impurities and product are extracted at various stages downstream. Caustic is the main constituent of the working fluid. There are two identical circuits, each with spare (stand-by) capacity incorporated either in parallel (as in the mills) or in series (as in the digester banks) or in both series and parallel. In some processes (e.g. the evaporator heat exchangers) there is no stand-by capacity. There is also some interstage storage in the main circuit — because of the involvement of precipitators — but its exploitation incurs a production-loss penalty. Although not part of the primary circuit, a number of sub-systems (e.g. the hydrate conveyor and kilns) spur off it . In each case these can be considered to be in series with the main circuit.

The refinery is operated continuously — it never comes off-line at plant level. The off-line maintenance work has to be scheduled by taking advantage of the extensive redundancy at unit level. The other important characteristic is that most of the maintenance is caused by time-dependent failure mechanisms such as wear, corrosion and fouling. A typical maintenance shutdown of a unit is estimated to involve a crew of nine tradesmen for about four days. The main exception to this is a kiln shutdown, which could involve up to twenty tradesmen for three weeks.

The existing unit life plans — e.g. for the bauxite mill — are based on fixed-time maintenance. A typical life plan would be as follows:

• Minor preventive work	Lubrication schedule.
	On-line inspection schedule
	(extremely limited).
	Service schedule (basic and
	not carried out well).
• Major preventive work	Major repair or overhaul at
	frequencies estimated from
	experience and plant
	history.

• Corrective guidelines for critical items.

Note that the major off-line work has been scheduled at fixed operating periods. The maintenance schedule has been arranged to spread the workload evenly across the year. The availability level is 92 per cent and Management consider that the direct cost of maintenance (men, spares, materials) is too high.

The problem

Senior Management want to improve the availability level of the plant to 95 per cent and at the same time to reduce maintenance costs by 10 per cent. They believe that the way to do this is to introduce a condition-based strategy.

The questions

(a) List the main factors that should be included in the maintenance objective. If the aim is to raise availability at the same time as reducing maintenance costs what are the factors that might inadvertently be neglected? How would you guard against this happening?

(b) Do you consider that the adoption of condition criteria for determining when units come off line would improve the maintenance strategy? How would such a policy affect the maintenance work load and organization?

(c) It has been suggested that a maintenance life plan for a plant (e.g. for an alumina refinery such as this) should be made up of a mix of the following procedures:

fixed-time maintenance;
condition-based maintenance;
operate-to-failure;
design-out maintenance.

If you were asked to formulate a strategy for a large complex plant how would you go about deciding what mix of procedures to employ?

(d) In general a maintenance strategy for a power station is based on fixed operating periods (of three years or so) between major shutdowns. Why do you think this is so? How does this approach affect the workload and the organization? What is the fundamental difference between the refinery strategy and a power station strategy?

Solution guidelines

Solutions to problems of this kind cannot be exact ones. The proposals below must be regarded not as *optimal* solutions but as guidelines to *good* solutions. Various of the points raised are open to debate.

Exercise 1

(a)   Condition-based maintenance has already been used for some time. The work therefore tends to come in on a random basis, but with at least a week's notice (or thereabouts) as a result of the inspection. In addition, the packets of work for each plant unit are small. This means that a resident workforce can handle this comfortably over normal (or extended) weekends.

(b)   If sufficient history is available frequencies can be determined, for fixed-time work, which will be effective in controlling reliability. However, it is unlikely that such an approach would be as cost-effective as a condition-based policy because there is always a degree of over-maintenance with fixed-time policies (you are playing the percentages).

(c)   The advice would be to continue with a condition-based policy and refine it as far as possible. This is not to say that some units might be better under a fixed-time regime and some might even be operated to failure (if the consequences of their failure were negligible).

(d)   Fixed-time maintenance (apart from calibration and, in some cases, cleaning) is not effective in controlling the reliability of solid state electronic items. It is best either to operate to failure on a planned basis — e.g. prepare for quick change of boards — or, if this is not acceptable because of the failure consequences, to consider designing out the problem (incorporating some redundancy, for example).

Exercise 2

(a)   The crusher circuit fails randomly so a fixed-time policy is not effective in controlling reliability.

(b)   The alternative policies are (i) design-out, (ii) condition-based, (iii) operate-to-failure.

Option (iii) is already in use and is proving too expensive. If option (i) is considered (as it must be) the causes of failure need to be identified and options considered for their elimination. This, however, is a *long-term* approach and the most cost-effective attack is likely to be the adoption of a condition-based policy. The information given is that the main causes of failure are wear, corrosion, or fouling. Therefore, for most items, monitoring techniques for predicting failure can probably be found and effort would need to be directed at the historically unreliable items. This might allow maintenance of the crusher circuit to move from corrective (following unexpected failure) to scheduled (albeit short-term) plus planned shutdowns.

Even in the case of unexpected failures, the monitored information should facilitate improved preparation and planning.

(c)   The main reason for ball mill shutdown maintenance is replacement of the lifters and liners. Their deterioration is time related and is statistically predictable so fixed-time replacement would be effective for controlling their reliability. It is not unlikely, however, that some form of condition monitoring might facilitate running the ball mills for the maximum time before the lifters and liners need replacing. In many cases this would take the running time past six months, but it might sometimes be as little as four months. However, if the inspection techniques gave an adequate planning lead time, the shutdown could be scheduled.

Exercise 3

(a)   The factors that could be neglected are standards of safety and plant condition (longevity). Corporate management must be made aware of the link between maintenance effort (and resources) and safety. The budget must take into account the longer term major maintenance work that influences equipment longevity.

(b)   The adoption of a condition-based approach could extend running time without reducing equipment reliability. This, however, assumes that a monitorable meaningful parameter can be found. If this is the case, condition-based policies might improve unit availabilities and also reduce maintenance costs. The downside of this could be that the workload might fluctuate erratically (perhaps with very large peaks). It would not be easy to co-ordinate maintenance work with production requirements or to use the 'common centralized maintenance resources' efficiently. If the workload varied erratically across such a large plant, the organization would need to be designed to match, i.e. resources would have to be *plant-flexible* or greater use would have to be made of contract labour. If the fixed-time work were largely retained, condition-based procedures might still be adopted for two reasons, namely:

●   to help predict the corrective work needed during shutdowns — this improves planning, and
●   to avoid unexpected failures.

(c)   Use a top-down bottom-up approach.

(d)   A base load power station shutdown might well take twelve weeks and employ as many as a thousand tradesmen. The date must therefore be fixed some considerable time ahead, to facilitate the necessary extensive planning and resourcing. The maintenance workload might have a peak/trough ratio (shutdown/normal) of up to 10:1, which would necessitate the employment, for the shutdown, of contract labour. The fundamental difference between the power station and refinery strategies is caused by the difference in the shape of the workloads. For the refinery, the work can be smoothed over the year and carried out by an internal labour force; for the power station, extensive use of contract labour — for resourcing the shutdown peak — will be necessary.

13
Reliability Centred Maintenance*

Introduction

In Chapters 8 and 9 a structured approach was developed for formulating the strategy for maintaining a large industrial plant, over its whole life. A fundamental feature of this approach, which has been developed by the author and his colleagues over a period of more than twenty years of consultancy and teaching in this field, is that it recognizes not only that the plant itself will probably be very large — comprising thousands of different units, each with different maintenance needs and interacting in ways that may be very complex — but that the maintenance strategy, *if it is to make its maximum contribution to profitability,* must take account of more than just the technical characteristics of the hardware, of the plant itself. Its formulation, from the outset (i.e. from the top down), must recognize such things as the plant's pattern of operation, the nature and availability of the maintenance resource, statutory safety requirements and so on. Most importantly of all it must take into account the pattern of demand for the plant's product, i.e. the relationship between the operation and the market. Hence the reason why the author has adopted the name 'Business Centred Maintenance' for this approach.

An alternative route to formulating maintenance strategy, one which been widely adopted, in major industries, is provided by 'Reliability Centred Maintenance (RCM)' and this chapter will now be devoted to outlining this. The reason for doing so is that its development predates that of BCM by a few years and key elements of the thinking behind the latter, which it is important to appreciate, were derived from it, as we shall see. It differs fundamentally, however, in that its main thrust is not so much towards maximizing the contribution of maintenance to the profitability of the organization as towards maximizing the technical performance of the plant itself. To quote from a leading consultant in its application:

> ... RCM is focused on the needs of the asset, not the shape of the organization[1]

The reason for this is that RCM was originally developed, as we shall see, for an industry, civil aviation (in the USA), which inevitably had to put the

---

* Chapter contributed by M. J. Harris, Honorary Fellow, University of Manchester School of Engineering.

safety (attained via the reliability) of its hardware first, before it could turn to thinking about economic matters.

History and basic philosophy of the RCM approach

The first thing to get quite clear is that RCM is *not* just a portmanteau term for those maintenance procedures which have been **scheduled** via some kind of Operational Research analysis using cost and reliability data: calculations, for example, of the optimum frequencies for given preventive tasks, proof tests, or inspections. In fact, such optimums exist only for rather infrequently occurring, rather special cases and searching for them is usually not all that productive (e.g. an optimum replacement frequency only exists if *both* (a) failure of the items concerned is due to a definite wear-out mechanism, and (b) the total cost — direct plus indirect — of a scheduled replacement is significantly less than the total cost of a failure replacement). The author has encountered, in a major company, the use of the term RCM to refer to just that kind of activity — and this is quite misleading. RCM is *much* more than that.

In RCM maintenance strategy is formulated via a structured framework of analysis aimed, in principle (but see later), at ensuring the attainment of a system's *inherent reliability*, i.e. the reliability that it was *designed* to attain. (It was a fundamental, starting-point assumption of the method's originators — F. Stanley Nowlan and Howard F. Heap of United Airlines in the USA — that this was the maximum level of reliability that *could* be attained.) The method incorporates several of the basic techniques of reliability engineering which were touched on in Chapters 5 and 6.

In the 1950s airlines in the USA (and elsewhere) were finding that as they introduced bigger aircraft, with more complex systems, their customary policy of periodically overhauling all systems — in the belief that each system would eventually deteriorate unless renewed — was generating an excessive, totally unachievable, workload. Indeed, they also suspected that safety and operational reliability were being threatened by unnecessary interference with satisfactorily running systems. A joint task force, comprising representatives of the civil airlines and of the regulatory Federal Aviation Authority, was therefore convened to study this problem. Significant among its many revealing findings, reported in 1960, was that over 80 per cent of aircraft items showed no evidence at all of age-related deterioration in function (i.e. no increase, with use, in hazard rate, see Chapter 5). The task force therefore concluded that:

(i)   scheduled overhaul had little effect on the reliability of complex items, unless wear-out was dominant (which was rarely the case);

(ii)   for many items there was no effective form of preventive maintenance.

During the 1960s further work carried out by a Maintenance Steering Group (MSG), comprising representatives of the FAA, the airlines and the manufacturers, showed that more efficient maintenance programmes could be developed using logical decision processes, progressively more refined formulations of which were eventually recommended in the handbooks MSG-1 (1968), MSG-2 (1970) and MSG-3 (1980). The first of these was used with great success to develop the maintenance programme for the then new Boeing-747, the later versions being applied equally successfully to the DC-10, A-300, Concorde, etc. The most significant outcome of this was probably the considerable shift from scheduled overhaul to condition-based maintenance (with great savings in maintenance man-hours and spares holding). The name 'Reliability Centred Maintenance' for the new approach was coined by its moving spirits, Nowlan and Heap, shortly after the publication of MSG-2, and was the title of the definitive handbook on the topic which they published at about that time[2]. Most of the variants of RCM that have since been developed for general industrial, as opposed to aviation, application (e.g. that of Moubray[3]) have, in fact been based on MSG-2.

Nowlan and Heap stated that the logic of RCM is based on three questions:

(i) How does a failure occur?
(ii) What are its consequences for safety or operability?
(iii) What good can preventive maintenance do?

and they further emphasized that, in RCM. . .

> the driving element in all maintenance decisions is not the *failure* of a given item, but the *consequence* of that failure for the equipment as a whole.

The RCM procedure

Figure 13.1 outlines first, in Steps 1–4, the basic structure of RCM analysis, namely:

(1) system definition and acquisition of operational and reliability information;

(2) identification of *maintenance significant items* (MSIs), i.e. items the failure of which would significantly threaten safety or increase cost (because of loss of production and/or high direct repair cost);

(3) for each MSI, determination of the significant failure modes, their likely causes, and whether they can be detected (and if they can be, the ways in which this might be done);

(4) for each significant failure mode, selection of the maintenance task, or tasks most appropriate for reducing its likelihood of occurrence or mitigating its consequences.

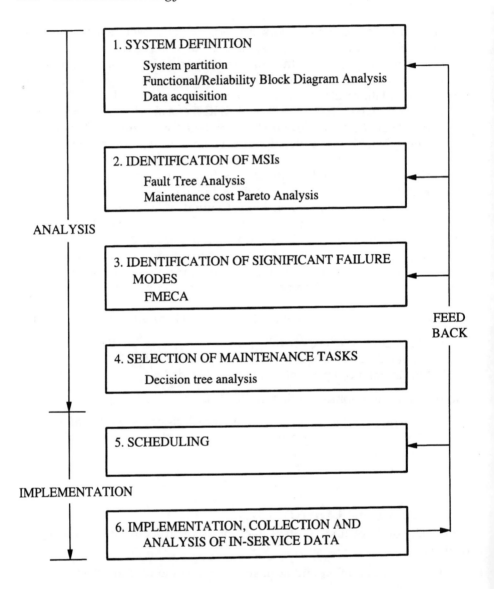

Figure 13.1. Basic structure of RCM

The analysis has then to be followed by:

(5)   the formation of the task list into a workable plant-wide schedule;
(6)   implementation of the schedule and sustained feedback of in-service data for periodic review and update.

Much of the analysis is a rational ordering of techniques that have long been well established and routine in reliability engineering (and were explained, or at least touched on, in Chapters 5 and 6). Step 1, for example, is basically system partitioning (separation into identifiable units) and reliability block diagram analysis. Although Step 2, identification of the MSIs, might well be accomplished just by reviewing history records, operator's logs and cost data to pick out the unreliable or maintenance-costly items (an activity facilitated by 'Pareto analysis', which is little more than sorting out the worst performers into a ranked list, e.g. of the 'top ten'), fault tree analysis might be needed if the plant is a complex one. Step 3 is nothing more than a failure mode, effect and criticality analysis (FMECA), a step-by-step procedure — based on documentation of the type illustrated in Table 13.1 — for the systematic evaluation of the failure effects and the criticality of potential failure modes in equipment and plant.

To these are added, in Step 4, the logical task-selection decision-tree which has been specially developed for RCM and is regarded by some as the kernel of the whole approach. In this, the question repeatedly posed, in order to filter out the various maintenance options, is:

> Is the task under consideration both APPLICABLE (Could it be done? Would it work?) and WORTHWHILE (Would its cost, direct and indirect, be less than that of just allowing the failure to occur?)?

The decision tree begins, however, with a *consequence analysis*, typically along the lines of the one displayed in Figure 13.2, which is based on the particular form recommended by Nowlan and Heap. By its means, significant failure modes are categorized according to their consequences, which, as regards their processing in the subsequent *task analysis* part of the tree (see Figure 13.3), are prioritized (in this version), as below:

(1)	*Hidden* (or *unrevealed*)	Increase risk from other failures (applies mostly to non-fail-safe protective equipment).
(2)	*Safety-related* or *Environmental*	Threaten life, health or environment.
(3)	*Operational*	Threaten output, or quality of service.
(4)	*Non-operational*	Incur only *direct* cost of repair.

Having been categorized by consequence in the upper part of the task selection tree, each failure mode is then subject, in the lower part, to a decision logic along the lines of Figure 13.3. This leads to identification of an appropriate maintenance policy or, if none can be found, to the suggestion that re-design be considered.

Figure 13.3 highlights the branch of the tree that is followed for a failure with *safety* consequences; for the other categories of failure the final, or lowest, questions are different, as indicated (e.g. for a hidden failure a failure-finding task, or proof test, should be considered before turning to possible re-design).

Table 13.1. Extract from a typical FMECA worksheet

Lubrication system
Ref. Drawing – XYZ 123
Operational state – Normal

Date:
Sheet 1 of n
Originator:
Approved:

Item identity/ description	Function	Failure mode	Failure cause	Failure detection method	Failure effect		Compensating provisions	Severity	Loss frequency $(f/10^6 h)$	Data source	Remarks
					Local effect	System effect					
22.2 Oil heater	Maintain lube oil temperature	22.2/1 Heater temperature unit failure	Open circuit	Oil temperature gauge	Violent foaming of oil during start-up	Low lube oil temperature. Fluctuating oil pressure	Pre start-up checks include oil temperature readings	2	29.0	Reliability Data Handbook	Heater maintains lube oil temperature during S/D
		22.2/2 External leak	Seal/ flange leak	Visual level gauge on oil reservoir	Loss of lube oil from system	Bearing or seal failure on drive unit	Bearing temperature high alarm. Automatic drive unit S/D on HH alarm	2	13.0		

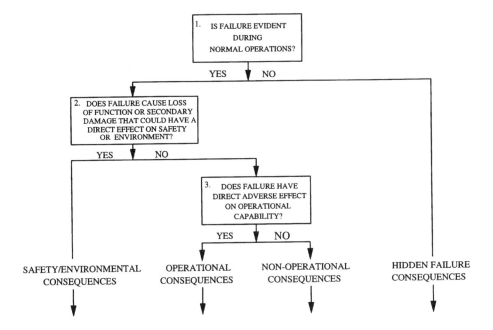

Figure 13.2. Consequence analysis

Note that the effect of the decision tree is to rank the task options in a definite order of preference. The analyst is directed to look *first* for a condition-based task, only *secondly* for a restoration (repair) task, and so forth. Implicit in the method is therefore the assumption that, where both of these options are viable, the condition-based one will be the more appropriate (and likewise the repair option as compared with replacement, etc.). Now, while this may well be true in most practical cases it is by no means obvious that it will be in all of them. In some forms of the decision tree that have been developed the analyst — before making his final selection — is, in fact, directed to finish with a comparison of all the types of tasks that have been identified as viable.

The example decision trees shown here have been derived from those presented by Nowlan and Heap in their handbook and are largely those of the MSG-2 stage of their work. Many other variants have been developed to meet the requirements of new safety and environmental legislation or of particular technologies. Moubray, for example, adds a branch into the consequence analysis stage to take separate account of environmental consequences[3]. Sandtorv and Rausand, in an interesting review of the application of RCM in the offshore oil and gas extraction sector, presented a quite different decision tree logic that took account of that industry's operating practices[4].

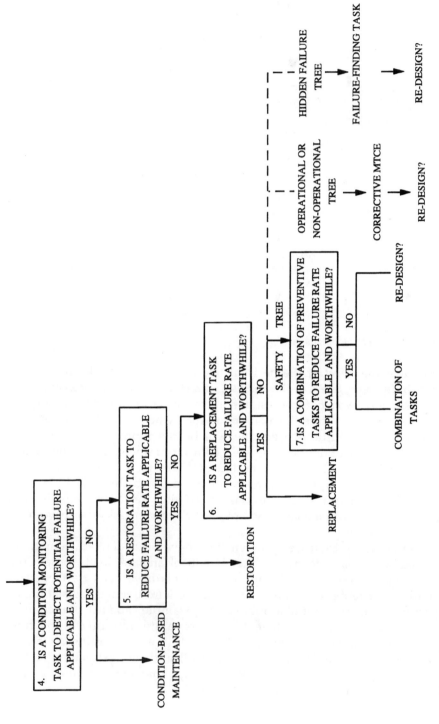

Figure 13.3. Task analysis

RCM in civil aviation

The benefits of RCM in its industry of origin have been outlined by Moubray[3]. For example, before its introduction the initial maintenance programme for the Douglas DC-8 specified scheduled overhaul of 339 items; that of the later DC-10, which was based on MSG-2, of only seven items, an improvement which led, among other gains, to a reduction in the spares inventory of more than 50 per cent. In addition to such organizational and economic gains, the resulting better understanding of failure processes has improved reliability by enabling preventive tasks to be directed at specific evidence of potential failures.

Various factors have contributed to this success:

- The RCM programme has had one clear objective, safety and technology driven, namely achievement, as mentioned earlier, of 'the inherent reliability capabilities' of the equipment concerned (and Nowlan and Heap also added 'and to do so at minimum cost').
- Aircraft systems are clearly specified and standardized, with much system redundancy. Also, they all have to fulfil a similar, mission-based, duty (for any one design, at least). The collection of data on reliability, availability and maintainability can therefore be relatively rapid.
- Much basic benchmark information for initial, design stage, RCM assessment can be 'borrowed' from the history of already functioning similar systems.
- As in other high technology sectors (e.g. nuclear power) there is a firmly established prescriptive culture and hence an acceptance of strategy directives produced by specialists.
- The work has been 'zero-based' — i.e. applied at the design stage with little reference to resource constraints, provided it is effective in controlling reliability.
- Last, but by no means least, the considerable cost of the programme can be spread over the total fleet.

In industry at large, however, the situation may be very different.

RCM in industry

Over the last ten years or so RCM analysis — broadly of the kind that has been described but, as explained, with occasional modification to suit a particular technology — has been tried in many different industries. The author and various of his colleagues have studied its application in a wide selection of these; viz. food processing, pharmaceuticals, offshore extraction, petrochemicals, steel production, metal forming, automobile manufacture, fossil-fired power generation, hydro and nuclear power generation.

We have found that, with notable exceptions — where RCM has been demonstrably beneficial and has become part of the company's culture — the success attained has not been of the level achieved in aviation. Indeed, in some cases RCM has been tried and abandoned, the company concluding that, for little and slow return, it demanded much greater resources of time and manpower than they had anticipated. There are several reasons for this, most of which arise because the maintenance management problem in these industries is, in many aspects, fundamentally different from that in the aviation business.

Although 'achieving the inherent reliability' is always adopted as the *de jure* objective of all RCM analyses and implementations, the *de facto* objective for most industrial plant (i.e. in the power, process, manufacturing and extractive sectors) may be rather different. More often than not, the plant concerned will already have been operating for some time, many years even. The level and mix of resources will have been set by custom and usage but will be perceived as being in need of reduction and rationalization. So the CM exercise will not be 'zero-based' and the dominant maintenance management objectives will be financial rather than technological. The aim will be to decrease maintenance costs — by justifying moving to contract rather than in-house arrangements, for example.

A typical industrial installation will often be a unique design assembled to meet a wide range of output requirements. There will be few standard systems. Collection of the necessary information on plant design and operation, and of data — whether generic or experiential — on reliability and maintainability will therefore be no small task and the high cost of the whole RCM exercise cannot be shared among similar plant.

Operation will frequently be via a single stream of diverse units run either on a batch basis, or continuously between major (often statutory) overhauls, features which may dominate maintenance decision-making, e.g. there may be clear windows of opportunity for preventive work at zero indirect cost. In such circumstances there may be relatively little that can be gained from a costly RCM exercise. A salutary illustration of this was recently encountered by the author. A pharmaceutical company had embarked on RCM in order to improve the generally poor availability of their plant. Although some small gains did indeed result, these were achieved only by putting a disproportionate effort, in time and manpower, into the study. Closer examination revealed that the greater part of the company's operation involved batch processing (with product changes) and that the dominant contribution, by far, to extended downtime was likely to be caused by delays in washdown and changeover. Effort devoted to improving the planning of these latter activities would clearly have been far more cost-beneficial, in the first instance, than the RCM exercise.

Unless special steps are taken (see later) there can be such a long delay between launching the RCM exercise and implementing its recommendations (intervals of two years or more have been recorded) that the latter can be overtaken by other changes — in plant, operating policy, sales requirement, and so on. A significant factor here is that translating the results of the RCM analysis into a workable maintenance schedule (Stage 5 of Figure 13.1) is invariably just as complex and demanding a task as the RCM analysis itself (see Chapter 9 for the TBDU approach to scheduling).

As explained, a number of the applications of RCM which have been studied by the author have, however, been successful, and this can be attributed to several features which they tended to have in common.

First, although achieving inherent reliability was always the notional working objective of the exercise, as it must be, other achievable objectives were identified, clearly defined and established as being primary, *and were not part of a hidden agenda.* For example, several companies had taken their main objective to be an educational one, to make Production and Maintenance aware of each other's needs and problems as part of fostering a self-managing, culture. Such an aim is usually very achievable because RCM, in most industries, is almost always undertaken not prescriptively, i.e. by specialist analysts issuing directives, but co-operatively and in-house, by facilitator-led operator-maintainer teams — an arrangement which can stimulate information exchange (and also, incidentally, the flow of previously unavailable reliability and maintainability data).

Secondly, a limited pilot exercise was invariably undertaken — checking preconceptions regarding the time and resources needed, revealing potential difficulties, and undertaken on a critical and representative sub-system. (This last point is an important one. Enthusiasm for, and commitment to, the study has been vitiated when an unimportant area of plant has been misguidedly selected.)

Thirdly, although when applied successfully to *new* plant RCM was carried out across the board — as an integral part of the design process — when applied to *existing* plant steps were taken to ensure that the analysis was only undertaken on those units where there were clear economic or safety benefits to be gained, i.e. on units critical to overall plant availability (identified by, say, reliability block diagram analysis, see Chapter 6) or on units exhibiting disproportionately high maintenance cost. In a study on an offshore oil and gas extraction platform, for example, the whole operation was analysed into just over a hundred sub-systems. It was then found that just 24 of these accounted for over 80 per cent of total maintenance man-hours expended (a classic case of Pareto's 'Law of Maldistribution'). Furthermore, the maintenance regime for half of these last 24 was already dictated by either legislative or code-of-practice requirements and could not easily have been changed. The RCM exercise was therefore confined to the remaining dozen sub-systems, which accounted for approximately 50 per cent of the man-hours. A halving of the predicted workload for these was achieved —

giving a reduction in the total expected maintenance workload for the platform of rather more than 25 per cent.

Fourthly, in the case of operating plant the exercise was organized in a way which specifically facilitated rapid implementation of its recommendations. At one plant, for example, all the RCM team members worked full time; schedules were entered on the computerized maintenance management system as they were compiled and the group did not break up until all the schedules were complete.

The benefits of RCM

Where it has been successful the benefits of RCM in general industry have been much the same as those claimed in the aviation (and military) sectors:

(i) *Traceability*. In the long term, the most important of the virtues of RCM. All maintenance policy decisions — and the information, assumptions and reasoning that led to those decisions — are fully documented. In the light of this, subsequent plant reliability can be periodically audited, maintenance experience reviewed and strategy updated (where necessary) on a rational basis.

(ii) *Cost saving*. As with aviation the overall maintenance workload is reduced, due to a general shift away from time-based or usage-based preventive work (such as regular major overhauls) and towards condition-based work — with a consequent reduction in spares holding.

(iii) *Rationalization*. By identifying unnecessary preventive work, unachievable, *and therefore uncontrollable*, maintenance workload is eliminated. In one section of a food plant, for example, the total *scheduled* preventive workload before the introduction of RCM was 25 000 man-hours — of which, typically, only 12 000 were completed, with no guarantee, of course, that it had all been directed at the more needful work. Under the RCM-determined regime it was established that the really necessary preventive work actually amounted to an achievable 12 000 man-hours.

(iv) *Plant improvement*. Re-design eliminates recurrent failures or poor maintainabilities.

(v) *Education*. The whole exercise raises the workforce's overall level of skill and technical knowledge. As mentioned earlier, this is a natural consequence of the operation of the facilitator-led study teams. Also, the actual existence of an RCM regime will itself tend to attract better-skilled personnel into maintenance.

RCM and the TDBU approach

If the ideas that underlie RCM are to be more widely accepted in industry, and the benefits of its rational approach fully realized, two fundamental aspects must be addressed:

(i) its strongly 'asset-centred' approach which, as explained at the start of this chapter, takes insufficient account of resource, operating pattern and market factors;

(ii) the resource-hungry nature of its implementation (2000 man-hours of team effort can be needed for even a quite modest project).

A consequence of the first of these is that operational and resource considerations are only fully addressed at the later, task selection, stage of the analysis and then only separately for each task, i.e. each time the question *Is the task worthwhile?* is asked. It is therefore no great surprise that the analysis is rather inefficient, i.e. resource-hungry (aspect (ii) above) if applied in its full form to a typical industrial plant where, as has been explained in Chapters 8 and 9, almost every aspect of maintenance strategy formulation will probably be dominated by plant-wide considerations, such as operational availability for maintenance.

The TDBU approach eases the above difficulties by providing an initial 'broad-brush' reliability, availability, maintainability and safety (RAMS) assessment of the plant concerned. An assessment which relatively quickly identifies firstly the operationally determined maintenance constraints and opportunities and then the maintenance-significant items. The in-detail core of RCM analysis, FMEA identification of significant failure modes followed by decision-tree selection of appropriate tasks for reducing their incidence or mitigating their effect (Stages 3 and 4 of Figure 12.1) can then be restricted to the most critical of the MSIs — i.e. employed if need be as part of Step 2, Stage b(iii) of Table 9.1. The TDBU approach has the additional advantage of providing a structure for systematically incorporating all the task recommendations, for non-critical as well as critical items, into a workable schedule for the whole plant.

References

1. Geraghty, T., Achieving cost effective maintenance through the integration of condition monitoring, Reliability Centred Maintenance and Total Productive Maintenance. *Maintenance*, **11**(1), 26–30, 1996.

2. Nowlan, F. S. and Heap, H., *Reliability Centered Maintenance*, National Technical Information Service, Springfield, Virginia, 1978.

3. Moubray, J., *Reliability Centred Maintenance*, Butterworth-Heinemann, Oxford, 1997.

4.   Sandtorv, H. and Rausand, M., Closing the loop between design reliability and operational reliability. *Maintenance*, 6(1), 13–21, 1991.

14
Total Productive Maintenance – its uses and limitations

Introduction

One of the major trends in European industry is the adoption of the Japanese technique (and underlying philosophy) of Total Productive Maintenance (TPM). Currently, its main users are the motor manufacturers, although it is beginning to be adopted elsewhere. This development is of such significance that the devotion of a chapter to its review is more than justified.

What is total productive maintenance?

This question is not easy to answer since every company seems to have its own interpretation. It is a technique which has been developed by Japanese manufacturing industry in order to provide both effective and efficient (and hence *productive*) maintenance in response to the needs of Just-In-Time (JIT) manufacturing and total quality management (TQM). Indeed, it has been said by one of its originators that JIT and TQM are just not possible without TPM. It was recently introduced in a conference as follows:

> There is nothing earth-shattering about TPM. It is a sub-set of *genba kanri* (workshop management), using a people-oriented approach to resolve maintenance and reliability problems at source.

A more formal definition and concept was given by Suzuki[1] (see Table 14.1 and Figure 14.1).

Table 14.1. TPM

1. Is aimed at maximizing equipment effectiveness — by optimizing equipment availability, performance, efficiency and product quality
2. Establishes a maintenance strategy (level and type of preventive maintenance) for the life of the equipment
3. Covers all departments — such as Planning, Production and Maintenance
4. Involves all staff — from top management down to shop floor
5. Promotes improved maintenance through small-group autonomous activities

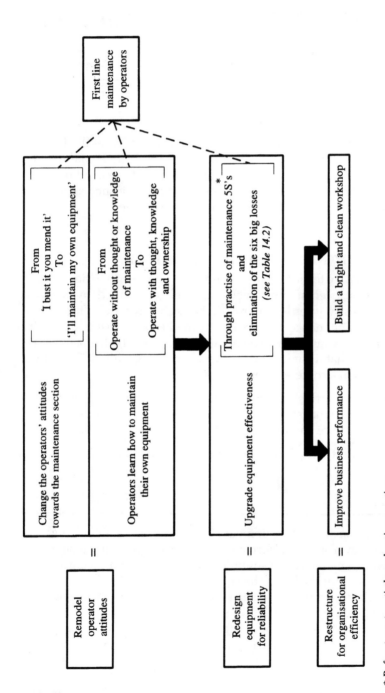

First line maintenance by operators

Change the operators' attitudes towards the maintenance section

From
'I bust it you mend it'
To
'I'll maintain my own equipment'

From
Operate without thought or knowledge of maintenance
To
Operate with thought, knowledge and ownership

Operators learn how to maintain their own equipment

Through practise of maintenance 5S's and elimination of the six big losses
(*see Table 14.2*) *

Upgrade equipment effectiveness

Build a bright and clean workshop

Improve business performance

=

Remodel operator attitudes

=

Redesign equipment for reliability

=

Restructure for organisational efficiency

* *Refers to systematic housekeeping practices -
seiri (orderliness), seiton (tidiness),
seiso (purity), seiketsu (cleanliness),
and shitsuke (discipline)*

Figure 14.1. TPM concept

The technique first surfaced in Japan in their manufacturing industry in the early 1970s — the first application being carried out at Nippondenso. It is now used throughout that country but is most strongly represented in the manufacturing sector (especially in Toyota-based companies). More recently it is being introduced, in a modified form, in their process industries[2]. Over the last few years it has been developed and implemented in some of the larger European companies — in both the manufacturing and the process sectors[3, 4, 5].

An early case study

The author's first contact with TPM was in 1977. The then Japanese Institute of Plant Engineering (JIPE) sent a maintenance management study tour to Europe. As part of this the group spent a few days at Manchester University exchanging views on maintenance management and, among other things, presented an explanatory case study of the ideas of TPM, based on the experience of the Toyoda Gosei Company Ltd, medium-sized suppliers of plastic injection and rubber mouldings to the car industry[6].

In the early 1970s the company was expanding rapidly, had neglected preventive maintenance, and was in the classic 'maintain it when it fails' situation, which is expensive in downtime and which engenders ineffective use of resources. In order to improve plant availability, product quality and resource utilization, the management decided to use TPM.

In order to incorporate the ideas of TPM into its existing organization the management used a 'small group circle' approach (see Figure 14.2). The TPM promotional activities were administered via a TPM promotions committee. The first step was to form a corporate TPM committee — which would decide on maintenance objectives and strategy — and departmental TPM committees — which would interact with the corporate committee and the voluntary small group circles of the shop floor. (This type of small group activity is a major feature of Japanese organizational culture. At one time the Toyota Motor Company, for example, had over four thousand such circles in operation.) Each of the committees and the circles had a membership which cut across departmental boundaries.

The committees suggested the aims and themes of the circles and also acted in supporting roles. Each circle (or sub-circle) appointed its own leader (who would then be a member of a higher group) and established its aims (within the theme set for it). The group was expected to find ways of achieving these aims and was given help and support as necessary.

One of the first conclusions of the senior committee was that the maintenance department should be more closely linked with production. This was carried out in two stages, as shown in Figure 14.3. The maintenance department was responsible for procurement of new equipment, setting technical standards and maintenance policy, and overall maintenance planning. The production manager was given the responsibility for the

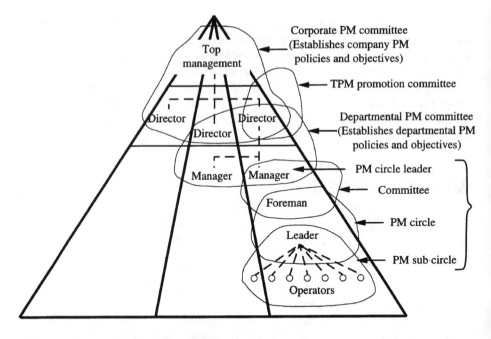

Figure 14.2. A system for promoting TPM within an existing organization

production and maintenance of his plant and had the maintenance foreman reporting directly to him.

Within this new plant-oriented maintenance organization the most important change was the creation of a new role for the plant operators, who were now expected to 'maintain normal operating conditions of machinery'. This meant that they had to operate the machinery, carry out inspection and cleaning routines, perform simple maintenance tasks and assist tradesmen as required. This necessitated expenditure of considerable effort to upgrade the operators' understanding of their machines and their maintenance 'know-how'. The small group circles were successfully used for this training activity and also for the promotion of a closer relationship between maintenance tradesmen and plant operators.

The TPM committee introduced a new maintenance policy, the thrust of which was based on the following:

- Mandatory daily and weekly inspections carried out by the plant operators (considerable effort was put into improving the plant for ease of condition monitoring).
- Improved corrective maintenance techniques (considerable effort was put into upgrading the tradesmen's maintenance know-how).
- Identification and correction of those plant abnormalities that caused low availability or high maintenance costs or poor quality, and feedback of such information to design for plant modification.

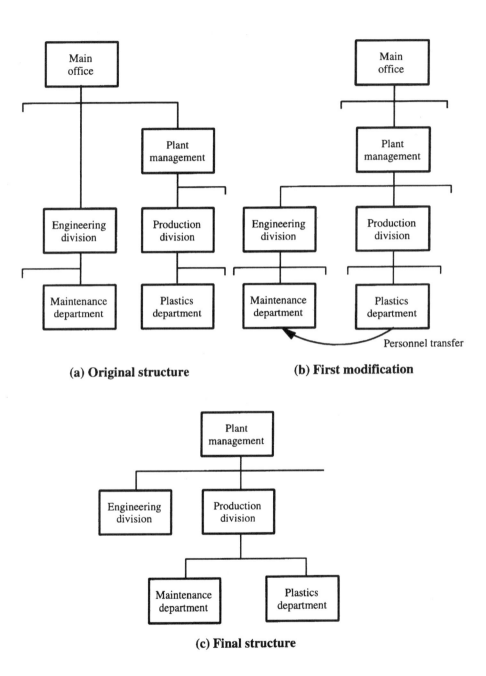

(a) **Original structure**

(b) **First modification**

(c) **Final structure**

Figure 14.3. Organizational modification for TPM

The TPM committee also emphasized the most important change of all — *that the new plan was to be carried out through the positive participation of all concerned*. The following example illustrates that such co-operation was indeed achieved.

A suggestion from a small circle, for reducing die-replacement time on moulding machines was implemented as a joint project by engineers from the die department, maintenance department and production division. Over a period of two years this reduced the replacement time from 49 minutes to 40 seconds.

The success of the efforts of the management and workers of Toyoda Gosei will be appreciated from the fact that, over a period of two years, the failure rate fell to 25 per cent of its original level, see Figure 14.4.

This impressive case study, and the accompanying discussion, revealed that the main concepts, principles and characteristics of the TPM technique were as follows:

- The company decided at corporate level to revolutionize its traditional maintenance strategy and practice.
- The industrial relations environment allowed the company to make such a change.
- The small-group activity was an essential part of the technique. In this case it was used in particular for design-out-maintenance (in the search for 'zero failures').
- Operators were given autonomous responsibility for maintenance and were given appropriate training.
- Equipment effectiveness was defined and downtime categories were identified.

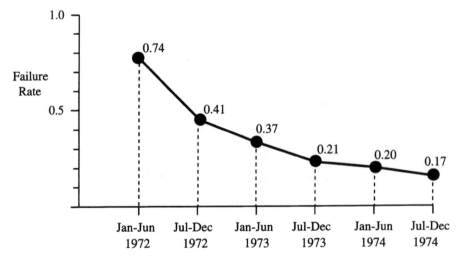

Figure 14.4. Reduction in failure rate

At this stage in its development (1977) the author considered that the main advantages of TPM over the then UK maintenance practice appeared to lie in the area of human factors rather than in systems or strategy.

Fundamentals of TPM

Over the last fifteen years the JIPE has been involved in promoting and consulting on TPM in Japan and world-wide. As a result, the technique has developed further and a number of JIPE-based books, and related papers, explaining its philosophy and application have been published[2, 7]. These, and various direct discussions with the JIPE, have led the author to an interpretation of the fundamentals of TPM as outlined below.

The basic philosophy[7]. Is to improve the way maintenance is carried out by improving its corporate image. Radical improvements based on new ideas and concepts are needed and are introduced by breaking away from past traditions and practices. One of the most important characteristics of TPM is that it must be accepted by the company as a whole — it is a total company-wide maintenance philosophy.

The maintenance objective or goal. Is not stated directly. However it is implied that the main aim is to maximize overall equipment effectiveness (OEE), where:

$$OEE = Availability \times Performance\ rate \times Quality\ rate$$

This is accomplished via the elimination of the six major losses (see Table 14.2) — reduction of downtime losses increasing availability, reduction of speed losses increasing performance rate and reduction of defect losses increasing quality rate. Suzuki[2] goes further and implies that the effects obtained by using TPM (and therefore the objectives aimed for) are those indicated in Table 14.3.

The maintenance policy. Can be considered to be made up of the following interrelated elements:

- Improving the effectiveness of the plant via an analysis of the OEE of each piece of equipment. In each case attempting to eliminate the six major losses and aiming for zero defects.
- Establishing a 'terotechnological' system with emphasis on the procurement, design and installation phases so as to ensure minimum maintenance life-cycle costs.
- Establishing, for existing and new equipment, a cost-effective maintenance life plan. This should include maintenance, spares holding and documentation policies.

Maintenance organizational characteristics. No one particular design of administrative structure seems to be recommended — in spite of the changes indicated in the introductory example (see Figure 14.3). However, the

Table 14.2. The six main losses

Downtime losses

1. Failures: losses caused by unexpected breakdowns
2. Set-up and adjustments: losses due to actions such as exchanging dies in press and plastic injection machines

---

Speed losses

3. Idling and minor stoppages: losses caused by the operation of sensors and by blockages of work on chutes
4. Reduced speed: losses caused by the discrepancies between designed speed and actual speed of the equipment

---

Defect losses

5. Defects in process: the production of defects and the reworking of defects
6. Reduced yield: losses that occur between the start up of a machine and stable production

---

following major organizational characteristics are an essential part of the technique.

- The establishment of company-directed small teams of operators who also carry out first-level maintenance activities — the so-called *autonomous maintenance teams*. An essential characteristic of these is the sense of plant ownership for their own area. The maintenance workers are also structured into groups of twenty and then into smaller teams (of up to seven), each with a leader.
- The use of small group activities — superimposed on the existing structure to promote, set up and monitor the use of TPM within the company (see Figure 14.2). This facilitates the top-down promotion of company TPM activities as well as the bottom-up generation of ideas for the shop floor group's activities.
- The mounting of a major effort in education and training. The small teams of operators, for example, go through the training steps indicated in Table 14.4, the last of which is aimed at engendering continuous improvement (or *kaizen*), i.e. the groups are encouraged to look for ways of assisting the maintenance teams or engineers in their pursuit of zero defects. Each of the groups also sets its own objectives and targets and monitors and records these results on an activity board for all to see.

Table 14.3. Some of the results of applying TPM

Measurable improvements
(A, B, C etc., refer to Company A, Company B etc.)

Productivity		Quality	
Total efficiency of equipment :	97% (D) 92% (N)	Fraction of defectives :	60% decrease (A) 90% decrease (T)
Labour productivity :	2.2 times (A) 1.7 times (I)	Number of claims :	Zero (T) 1/9 (D)
Number of troubles :	1/20 (I) 1/15 (S)	Cost for work-in-process :	1/4 (Y) 1/2 (N)
Productivity of added value :	1.5 times (T)	Lot out :	Zero (T)

Cost		Delivery	
Cost decrease :	50% (R) 30% (A)	Inventory decrease :	40% decrease (T) 50% decrease (S)
Energy saving :	1/2 (I and D)	Leadtime :	1/2 (A)
Maintenance cost :	40% decrease (A)	Turnover rate :	1.3 times (I and T)
	60% decrease (K)	Direct shipment rate :	60% (T)
Manpower saving :	1/2 (I)		

Safety		Morale	
Holiday accidents :	Zero	Number of patents :	37 (I) 28 (0)
Accidents with no rest :	Zero		
Labour accidents :	Zero	Number of suggestions :	30 times (I)
		for improvement	5 times (T)
		Number of national :	7 times (N)
		qualified experts	2 times (N)

'Invisible' improvements

(1)   Human resources
*   Confidence that we can do what was thought not possible
*   Staff take care of their equipment through self-maintenance activity
*   Leaders upgraded by practicing 'progress and harmony' again and again through the positive activities of group leaders

(2)   Equipment
*   Safe operation with decreased breakdowns and fewer unexpected small line stoppages
*   Realization of 'Streaming Factory' with less work-in-process and improved physical distribution through progress toward line production

(3)   Management
*   Improvement of the objective achievement rate after CAPD is activated
*   Do, and report
*   'Competitive consciousness' between the staff of plants by sharing the same goal and activities of a company
*   'Sense of unity' among them by participating in and exchanging inspection visits

(4)   Company image
*   Improves the image of business partners and the group companies by getting them to know about the introduction of TPM, by word of mouth in the area and/or inspection visits

Table 14.4. The seven steps for establishing operator-maintenance groups

Step	Name	Contents of activities
1.	Initial clean-up	All-round clean-up of dust and dirt, centring on the equipment proper, and implementation of lubrication, and machine parts adjustment; the discovery and repair of malfunctions in equipment
2.	Measures against sources of outbreaks	Prevention of causes of dust and dirt and scattering, improve places which are difficult to clean and lubricate. Reduction of the time required for clean-up and lubrication
3.	Formulation of clean-up and lubrication standards	Formulation of behavioural standards so that it is possible to steadily sustain clean-up, lubrication and machine parts adjustment in a short period (necessary for indicating a time framework that can be used daily or periodically)
4.	Overall check-up	Training in check-up skills through check-up manuals; exposure and restoration of minor equipment defects through overall check-ups
5.	Autonomous check-up	Formulation and implementation of autonomous check-up sheets
6.	Orderliness and tidiness	Standardization of various types of on-the-job management items and complete systemization of upkeep management • Standards for physical distribution in the workplace • Standards for clean-up, check-ups and lubrication • Standardization of data records • Standardization of die management, jigs and tools
7.	All-out autonomous management	Development of corporate policies and goals and making improvement activities routine MTBF recording and analysis, and consequent equipment improvements

European applications by non-Japanese companies

Several non-Japanese companies have used the ideas and concepts of TPM in an attempt to improve maintenance performance.

The Volvo car assembly plant at Ghent has carried out a major reorganization which involves the establishment of small self-managing teams of operators responsible for quality, operation and first-line maintenance in individual production process areas. These teams have followed the standard TPM training steps indicated in Table 14.4, focusing on continuous improve-ment. The maintenance department carries out second and third-line maintenance with major emphasis on planned preventive maintenance programmes. Thus, TPM in this plant lays stress on the idea of small self-managing teams of operators. In addition, the whole re-organization was implemented via classic overlapping small group activities (see Figure 14.5). As a result, the plant has seen considerable improvements in production, quality and reliability. It appears that Volvo used their own management to design and undertake the implementation.

A process plant example of TPM is that of Hoechst, in France, who used the JIPE as consultants and advisers[5]. They concentrated on the following three areas:

(1) Improvement of the reliability and maintainability of new equipment — via company and manufacturer analysis and systems. Hoechst called this the AMEDEC procedure — making new investments reliable.

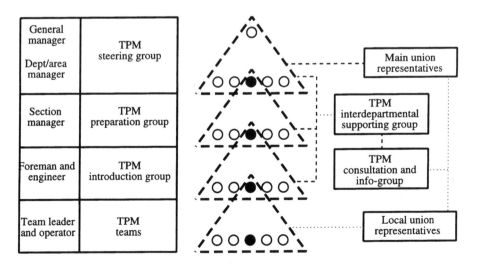

Figure 14.5. Overlapping small groups at Volvo

(2) Improvement of OEE — via the analysis of the six major losses for existing critical units. Hoechst called this the TRS technique. They showed, as an example, how it had improved production output of a critical unit by 30 per cent.

(3) The formation of small groups of operator-first-line-maintainers (they called this 'automaintenance'). Although considerable effort has been put into this area, through restructuring and training, Hoechst appear to be not yet convinced about its success.

TPM has also been used in a more conventional way in the steel company Usinor Sacilor[8] and in the car manufacturers Renault, both in France[3, 8]. Both companies are satisfied with the improvements in production, quality and reliability thus obtained.

TPM the Nissan (UK) way[9]. Nissan Manufacturing (UK) (NMUK) was established in 1984 as Nissan's foothold in the European market. By 1993 the total investment in NMUK was around £670 million, employing 3500 people, and producing some 200 000 cars per year. The plant was located on a green-field site in Sunderland and is a fully integrated car manufacturing facility (see Figure 14.6). NMUK has negotiated a single union agreement with the AEEU — about 28 per cent of the workforce having membership.

The resource structure outlined in Figure 14.7 shows that each major plant has its own tradeforce located in its own workshop, the operators

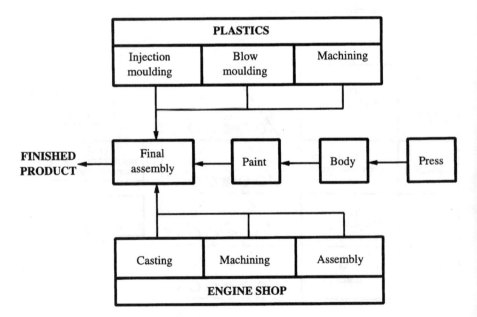

Figure 14.6. Plant layout, Nissan (UK)

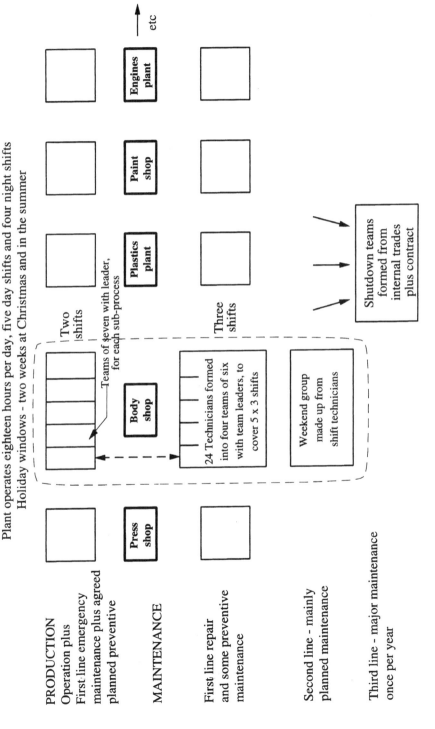

Figure 14.7. NMUK resource structure

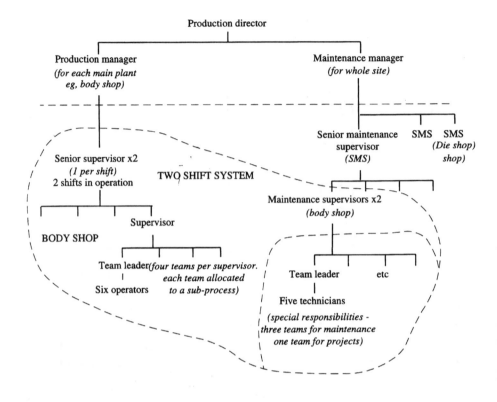

Figure 14.8. NMUK administrative structure

undertaking the first-line maintenance. The tradeforce work on shifts and back up the operators' first-line maintenance during normal running. In addition they carry out, where possible, limited second-line work during normal running and also weekend planned maintenance. Figure 14.8 shows that the administration is functionalized at the top — into Production, Engineering, Sales, etc. This structure is further functionalized, under the Production Director, into Production and Maintenance and remains thus divided down to shop floor level — in many respects a 'traditional' large structure comprising many departments and a number of levels of management.

Perhaps the key organizational characteristic is the way the first-line supervisors and shop floor are structured into groups. Figure 14.8 shows that both production and maintenance supervisors are responsible for groups of about twenty workers. In the case of production the groups are further divided into three teams, each with six or seven operators and a leader. The supervisor is responsible for a zone (a process area) and the team leader for a sub-process. Two maintenance supervisors (one of whom, per shift, links with Production) are responsible for the body shop. They have four teams, each with a leader, to cover fifteen shifts — Monday to Friday. The

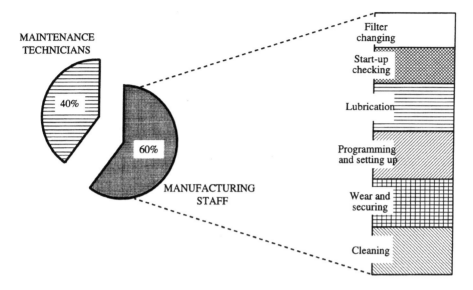

Figure 14.9. Division of maintenance activities

function and training of these groups and teams, and the systems employed, are essential to the Nissan way of management.

Each group has a degree of autonomy. The supervisor is the focal point for recruitment, training and solving industrial relations problems. The team leader helps to train the group and looks after day to day planning. There is considerable peer pressure within each group and there is undoubtedly a sense of plant ownership within each operation team. They carry out the maintenance tasks shown in Figure 14.9 and are also heavily involved in continuous improvement (*kaizen*) projects. Each group also puts considerable effort into the ideas of visual management, it being expected to produce and display regularly updated charts of the key group-performance indices — of attendance, quality, output and so forth. These are on display near where the group works and for all to see.

The maintenance teams carry out the planned preventive activities, respond to production first-line maintenance requests and train the operators in their TPM tasks. In addition, they get involved in group analyses of maintenance problems.

TPM is an essential part both of NMUK's general managerial philosophy (*genba kanri*) and of its procedures, features of which are as follows:

- Objectives are set at corporate level and translated down to the shop floor groups. Every employee has a copy of his or her objectives on a personalized card.
- *Kaizen* is not only a part of TPM but is promoted through small group activity at all levels of the organization.

- *Ad hoc* and company-led teamworking is used for ongoing vertical or horizontal communication and for the initiation and development of special projects.
- The 5Ss housekeeping approach (see Figure 14.1) is used to foster optimal working conditions on the shop floor.
- A major programme of education and training provides high quality personnel.
- Standard operating and maintenance procedures are established.
- Management control systems are adopted which, in the case of the maintenance department, include the following:

  - measurement of line operating ratio,
  - achievement reports,
  - audits of work standards,
  - review of preventive maintenance procedures.

Conclusions

During the last ten years the author has studied TPM implementations in a range of companies, some European and some Japanese owned. He has been asking the following questions:

What is TPM?
How is it different from other techniques?
Has it been used successfully, and if so why?

Regarding the first of these a response has been attempted in this chapter. The others are a little more difficult to answer.

How is it different? In terms of management procedures and maintenance systems there is nothing new.

- *The terotechnological approach.* Was pushed hard in the UK in the early 1970s and there were some particularly successful applications of the idea[6] (see also Chapter 2).
- *Maintenance strategy.* Many UK and European companies have an excellent approach — especially as regards condition-based programmes (see also Chapters 8 and 9).
- *First-line maintenance by operators and inter-trade flexibility.* Many Western examples could be given, notably Shell Chemicals, Carrington, which is a centre of excellence in this matter.
- *Self-empowered groups.* Zeneca (pharmaceuticals) and many others have successfully adopted this idea.
- *Small group activity.* Is based on the work of Rensis Likart[10] and has been applied in many Western organizations.
- *Continuous improvement.* For many years the author has advocated the approach to this that has been shown in Chapter 10. It has

been adopted by several Australian companies.
- *Computerized documentation systems*. In this area the USA and the UK are in a particularly strong position, having some of the most advanced and innovative systems.

So, to reiterate — in the systems area there is nothing new in TPM. Indeed, most of its concepts and systems have long been established in the USA and in Europe, various companies having employed a number of the above techniques in combination. Although this knowledge has been available to them, most European and USA companies cannot, however, be regarded as having *pro-active* maintenance departments. They usually function only *reactively* — because of the human factors problems of their conventional organizations.

In the UK the problem of vertical polarization — especially between the shop floor and the rest — is endemic and management rarely has the goodwill of the shop floor. This problem is often exacerbated by the divide between first-line supervision and professional engineering staff. In addition, there is almost always a horizontal polarization between the production and the maintenance departments — 'we bust it, you repair it'. Superimposed on this is the corporate attitude that maintenance is a 'fixed overhead'. This, in conjunction with a rigid budgeting and costing system makes for short-termism as far as maintenance decision-making is concerned. In such an environment it is therefore hardly surprising that most maintenance initiatives come from within the maintenance department and are unlikely to succeed fully because of the lack of support of other organizational and functional groups. Even when such initiatives come from corporate management they can fail because of the lack of a promotional procedure in a context of organizational resistance to change.

It appears that TPM succeeds not because of its strategy or systems or engineering techniques, but because of the following:

(i)   A belief by corporate management in the importance of maintenance and the realization that some resources have to be expended for long-term gain.

(ii) The use of the small circle approach, superimposed on the existing organization to initiate, and foster the acceptance of, the ideas of TPM. Or, as the Japanese say, 'it will not work without the participation of all concerned'. This approach was illustrated in Figures 14.2 and 14.5 and an example of it, for a large manufacturing company, is shown in Figure 14.10[2].

(iii) The traditional features of Japanese organization that tend to break down organizational polarization and create the ideal environment for TPM, i.e.

- overlapping groups to enhance horizontal and diagonal communication;

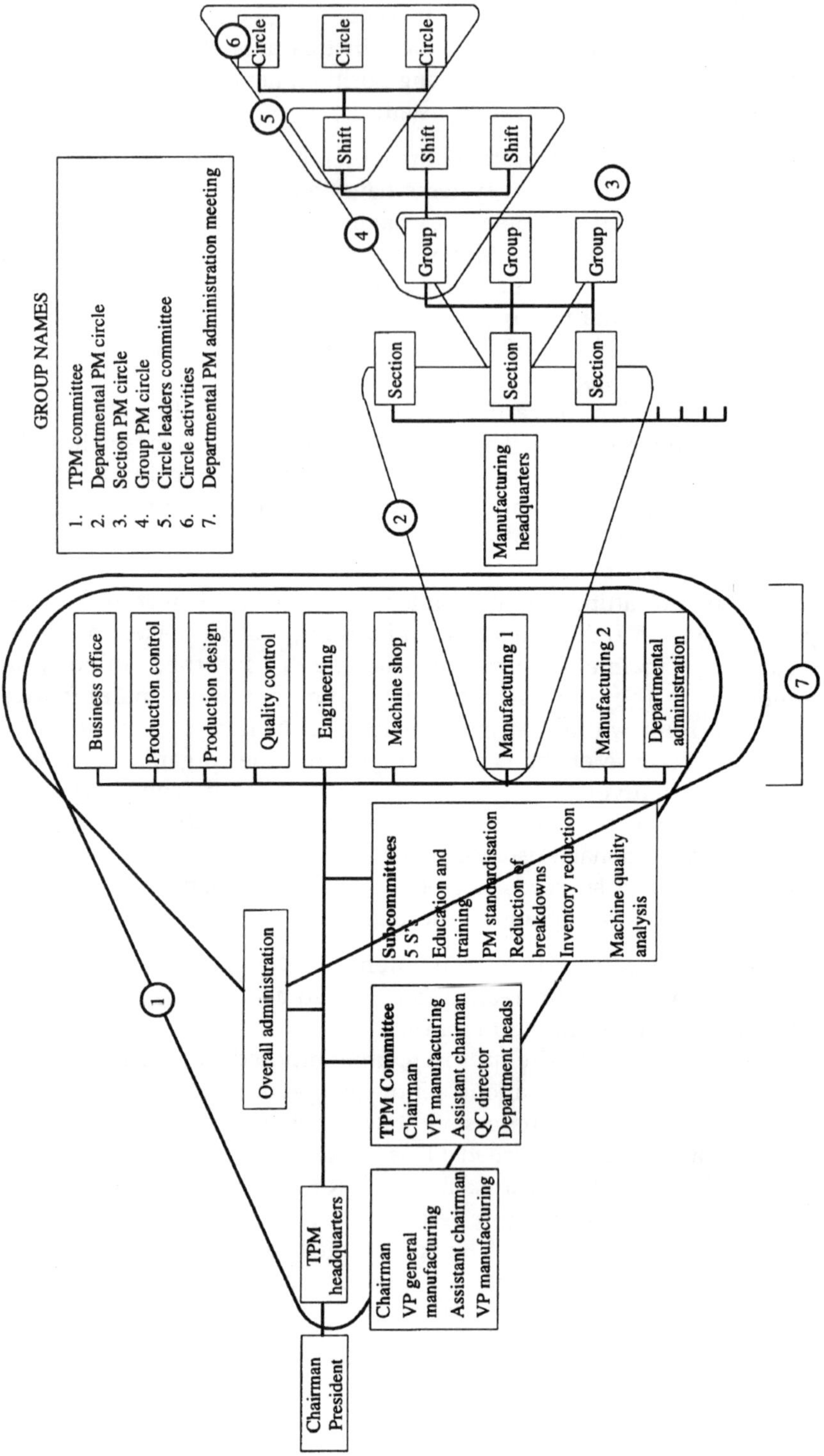
GROUP NAMES
1. TPM committee
2. Departmental PM circle
3. Section PM circle
4. Group PM circle
5. Circle leaders committee
6. Circle activities
7. Departmental PM administration meeting
Chairman
President
TPM
headquarters
Chairman
VP general
manufacturing
Assistant chairman
VP manufacturing
TPM Committee
Chairman
VP manufacturing
Assistant chairman
QC director
Department heads
Overall administration
Subcommittees
5 S's
Education and
training
PM standardisation
Reduction of
breakdowns
Inventory reduction
Machine quality
analysis
Business office
Production control
Production design
Quality control
Engineering
Machine shop
Manufacturing 1
Manufacturing 2
Departmental
administration
Manufacturing
headquarters
Section
Section
Section
Group
Group
Group
Shift
Shift
Shift
Circle
Circle
Circle
1
2
3
4
5
6
7

Figure 14.10. A TPM promotional structure

- empowered shop floor groups, further sub-divided into teams;
- an extraordinary level of 'company values indoctrination' and conventional training.

It is in the area of human factors management that we have most to learn from the Japanese. On his visits to Japan in 1978 and 1979 the author was impressed by many of the maintenance systems he was shown — including TPM. At that time he did not feel they could be applied in the UK because of that country's industrial environment. In addition, he felt that even if industrial relations improved, the UK culture was so different from the Japanese that the techniques would not be transferable. The experience of Nissan, Renault, Volvo and Hoechst shows that he was wrong.

References

1.   Suzuki. T., *New Directions for TPM*. Productivity Press, Cambridge, Massachusetts.
2.   Suzuki, T., New trends for TPM in Japan. *Total Productive Maintenance Conference*, MCE, Brussels, April 1992.
3.   Grossman, G., TPM at Renault. *Total Productive Maintenance Conference*, MCE, Brussels, April 1992.
4.   Poppe, W., Autonomous maintenance, autonomous quality. *Total Productive Maintenance Conference*, MCE, Brussels, April 1992.
5.   Bisson, F., Managing the production process. *Total Productive Maintenance Conference*, MCE, Brussels, April 1992.
6.   Kelly, A., *Maintenance Planning and Control*. Butterworths, UK, 1984.
7.   Nakajima, S., *TPM — Maximising Overall Equipment Effectiveness*. Paper published by the Japanese Institute of Plant Maintenance, Minato-Ku, Tokyo, Japan, (undated).
8.   Barbier, C., TPM in the steel industry. *Total Productive Maintenance Conference*, MCE, Brussels, April 1992.
9.   Fernie, A., MEng thesis, University of Manchester, 1992.
10.   Likart, R., *The Human Organization: its Management and Value*, McGraw-Hill, New York, 1967.

15
Conclusions

At present, there would appear to be three distinguishable approaches (see Figure 15.1) to the task of formulating maintenance strategy.

(i) Reliability Centred Maintenance (RCM)
(see Chapter 13)

As indicated in Figure 15.2 (a repeat, for convenience, of Figure 3.1) the main decision-making thrust of BCM, the approach that has been advocated in this book, stems directly from the business objectives. With RCM, however, this is not the case. RCM is more concerned with maximizing the technical performance of the plant — by addressing the consequences of item failure — and concentrates its efforts around the formulation of unit life plans rather than on the organization and systems areas of maintenance. It has been found to be a useful vehicle for the training and education needed in a Total Productive Maintenance programme and can also be a powerful aid in the search for failure cause (i.e. it can be used as an integral part of reliability control, see Figure 10.1). In summary, it is an asset-based approach best suited for replicated high-technology plant where safety and environmental considerations are overriding (e.g. nuclear power units).

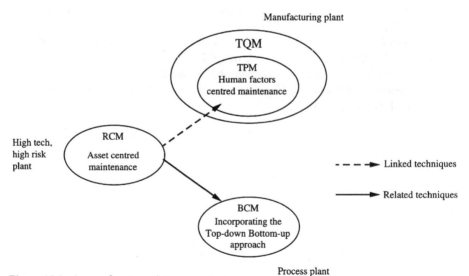

Figure 15.1. Approaches to maintenance strategy

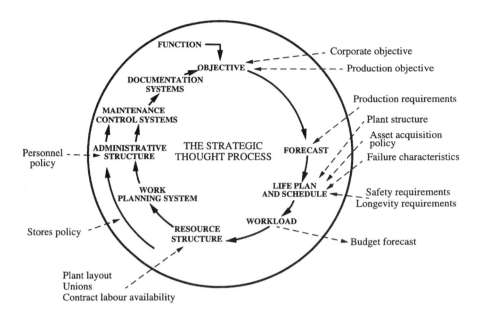

Figure 15.2. Business centred development of maintenance strategy

(ii) Total Productive Maintenance (TPM)
(see Chapter 14)

TPM covers all of the elements of maintenance management indicated in Figure 15.2 and at first sight might therefore be taken to offer a comprehensive approach to strategy formulation. It has evolved, however, out of the practices of Japanese manufacturing industry and, as a consequence of this, is presented more as a human factors oriented 'recipe' for setting up a maintenance department rather than as a generic approach to strategy formulation. It has had considerable success in a number of European and US manufacturing companies but in its fully fledged form it has not been %anything like as successful in the process sector. In summary, it is a human factors centred approach effective for manufacturing plant operating under a Total Quality Management (TQM) regime.

(iii) Business Centred Maintenance (BCM)

BCM provides a comprehensive and generic approach to the formulation of maintenance strategy. As indicated in Figure 15.2, it derives its impetus from the identified business objectives, which are translated via the top-down bottom-up (TDBU) analysis — an integral part of the approach — into user requirements for the plant units. The TDBU analysis, which provides a procedure for the detailed formulation of unit life plans and the plant maintenance schedule, makes use of many of the concepts and procedures (suitably modified for use in process plants) of RCM and also provides the link (via its estimation of the plant workload) between the life plans and plant schedule on the one hand and the organizational systems and documentation elements of maintenance management on the other. In summary, Business Centred Maintenance is a pragmatic approach particularly applicable to process-type plant.

This book has aimed to explain the overall methodology of BCM and the application of the TDBU approach to the setting of maintenance objectives and to the determination of unit maintenance life plans and the plant maintenance schedule. The review of BCM is completed in a companion book (see preface) which focuses on maintenance organizations and systems.

APPENDIX 1
Maintenance terminology

Definitions are given below of some of the more important terms used in this book. They are broadly in line with those given in the British Standards publication BS 3811:1984, but some have been significantly amended or extended by the author. The list does not include definitions already given clearly in the main text.

Maintenance
: The combination of all technical and associated administrative actions intended to retain an item in, or restore it to, a state in which it can perform its required function.

Preventive maintenance
: The maintenance carried out at pre-determined intervals, or corresponding to prescribed criteria, and intended to reduce the probability of failure or the performance degradation of an item. Preventive maintenance is planned and scheduled (or carried out on opportunity).

Condition-based maintenance
: The preventive maintenance initiated as a result of knowledge of the condition of an item derived from periodic, routine or continuous monitoring.

Condition monitoring
: The periodic, routine, or continuous measurement and interpretation of data to indicate the condition of an item.

Corrective maintenance
: The maintenance carried out after a failure has occurred and intended to restore an item to a state in which it can perform its required function. Corrective maintenance can be planned and scheduled.

Emergency maintenance
: The corrective maintenance which it is necessary to put in hand immediately to avoid serious consequences. Thus, emergency maintenance cannot be scheduled. In some cases, however, it can be planned for by ensuring that decision guidelines have been prepared and that necessary resources will be available.

Unit life plan
: The programme of preventive maintenance work to be carried out on a unit of plant unit over its entire life.

Preventive maintenance schedule
: A schedule of preventive maintenance work for the whole of a plant (or plant section). The schedule is a listing of jobs, with trades, against plant units and dates.

Maintenance window
: The opportunity to carry out off-line maintenance on a plant without incurring production loss. Windows can arise at plant, unit or item level.

On-line maintenance
: Maintenance which can be carried out while the plant or unit is in use (also called running maintenance).

Off-line maintenance
: Maintenance which can only be carried out when the plant or unit is not in use.

Appendix 2
In situ repair techniques

*(from a dissertation by Julia Gauntly,
Manchester University 1986)*

	Technique	Principles of operation	Application
1.	Inerting of flammable material storages	Foam generator used to inject inert gas	Makes safe for welding
2.	Welding and machining	Building up worn metal parts by welding until oversize and then machined back to size	Normally carried out in workshop but portable machines and welding equipment are available for on-site work (worn shafts, bearing housing gears, etc.)
3.	In situ machining	Full range of machine tools and hand tools available	Machine tools have emphasis on portability with special emphasis placed on devices to fasten the machine to the job. Limitless applications e.g. in situ grinding of rollers, machining the back face of a heat exchanger
4.	Flatness checking with monochromatic light and optical flat	An optical flat (flat piece of glass used as a reference) is placed on surface to be tested. Surface and flat are placed under a monochromatic light source. Interference fringes allow surface to be compared to optical flat	Used in conjunction with on-site machining operations such as grinding
5.	Alignment checking with lasers	Laser light is collimated (it propagates in narrow beams which have low divergence). A laser beam is emitted from a laser/detector unit mounted on the shaft of a stationary machine. It is aimed at a prism mounted on the shaft of the machine to be aligned. The beam is reflected back to the detector. The two shafts are rotated and the misalignment can be measured by the laser unit and corrected	Alignment of roller bearings in conveyor belt systems, drive shafts, fan motors, etc

6.	Laser cutting, welding, cleaning	A laser beam can be focused onto a small spot in order to give high energy densities of the order of those used for electron beam welding	Can be used for high quality precision welding and cutting. Not yet used extensively for in situ repair
7.	Laser gas absorption	This is a form of leak detection. It can be used to scan every area of a plant in seconds for leaks of a gas which absorbs radiation in the infra-red spectrum of a carbon dioxide laser. The system is based on a laser, mirrors and detector	Used to scan a plant for gas leaks after a major overhaul and start up, e.g. ammonia leak
8.	Leak sealing under pressure via sealing compounds	Manual or hydraulic injection of thermo-setting compounds into, or around, a leak. The application of heat, either externally or internally, from the contents of the pipe, e.g. steam, causes the compound to cure and seal the leak	Sealing flange leaks, heat exchanger joints, turbine joints, etc.
9.	Leak sealing under pressure (other methods) (a) Welding	Pinhole leaks can be closed by welding an ordinary nut around the leak. A bolt, with a sealing compound is then screwed into the nut	Pinhole leaks. Larger leaks can be tackled using a specially prepared plate rather than a nut
	(b) Clamping	A range of clamps can be purchased for sealing leaks in pipes	As above but do not require welding. Used for low pressure leaks
10.	Pipe lining	A terylene felt tube, impregnated with polyester resin, is inserted into an existing pipe and cured in situ	Used for the renewal of brick sewers, cast iron pipes, concrete culverts, etc.

11.	Pipe freezing	A method of isolating sections of pipe or plant (where valves are not available) by freezing the contents of the pipe, using dry ice or nitrogen	Routine maintenance of service pipes; extension of existing pipework systems
12.	Tube plugging	Used to seal off the leaking tubes in a tube bank by inserting a plug into the ends of the damaged tubes	Boilers and heat exchangers
13.	Explosive techniques		All kinds of boiler and heat exchanger tube repair
	(a) Expansion	Small explosive charges are detonated within the mouth of a heat exchanger tube to seal the tube–tube plate joint	
	(b) Welding	Similar to (a) but the conditions are arranged so that a weld is formed between tube and tube plate	
	(c) Loosening	The loosening of scored threads and other similar seizures	
14.	Hardfacing repair techniques	Covers the techniques listed below which are used to coat components with a surface which is best able to withstand the conditions encountered in service	
	(a) Thermal spraying	Flame spraying with wire used for thick coatings	Pump shafts
		Flame spraying with powder — used for small items	Impellors
		Plasma spraying with powder — used for spraying chromium and tungsten carbide	Fan blades, etc.

(b)	Flame plating	Particles of a metallic compound (tungsten carbide) are mixed with oxygen and acetylene in a 'gun' and then detonated. The metal is melted at high speed and 'sprayed on to the surface'	Fretting surface, gas turbine blades Worn shafts in gas compressors and steam turbines
(c)	Spray fusing	A two-stage process in which a coating is flame sprayed on to the workpiece, and then fused with an oxy-acetylene torch	Coating and building up worn pump pistons, sleeves, wear rings, etc.

(d)	Depositing	Depositing materials on to surfaces using welding techniques:
		
		    Oxy-acetylene rod and powder
		    Gas tungsten arc or argon
		    Metal arc
		    Plasma
			As for spraying techniques. Worn cutting edges and teeth on excavators, worn shafts and many other applications

15.	Brush plating	An electrolytic method of metallizing a surface without an electrolyte bath. The surface of the component is 'brushed' with an anode which is wrapped in an absorbent material (cotton wool) which has been dipped into an electrolyte	For coating worn surfaces or, depositing a corrosion resistant material, e.g. cobalt
16.	Hot tapping	A method of connecting branches to pipes which are under pressure and cannot be isolated. A branch is welded to the line to be hot tapped. A valve is fitted to the branch. A special drill is fitted to the valve. The valve is opened and the line is drilled. The valve is closed and drill removed. A new line is fitted to the branch	Used for connecting branches to main lines which are expensive to shut down and purge

17.	Cold tapping	Very similar to hot tapping but there is no welding. Instead of a conventional branch a 'tee clamp' is used which is clamped to the line	Used where welding would be dangerous
18.	On-line valve replacement	Used to change valves under pressure, (see Figure 8.10)	Valves can be removed without having to drain the system
19.	Thread inserts	Sleeves, usually with internal and external threads, which are used to replace damaged threads	Repair of damaged threads
20.	Metal stitching	Cold stitching a component which has fractured. Consists of drilling special apertures into both sides of the fracture and then peening matching keys into the apertures	Can be carried out on any metal over ¼" thick, e.g. machine foundations, gearbox castings, cylinder blocks. Most repairs can be carried out in situ
21.	Repair of floating tank roofs	Leaks are often caused by failure of welds and rivets on a roof under the stresses of its movement. Sections of plate are cut to fit over the leak and held in place by specially designed bolts and the edges sealed with a proprietary sealant	In situ technique for tanks storing all refinery products
22.	Repair of glass-lined vessels	A number of techniques, e.g. • Cement repairs • Tantalum plugs for pinhole leaks	Can be used for repairing in situ holes of a wide range of sizes
23.	Cold-forming materials	Generally consists of two or more components (liquids or putties) which are mixed together to form a uniform material. After mixing, the material is applied to the surface and is allowed to cure	

	(i)	Metal repairs — mixtures of metals and epoxy resins	Holes in pipes, scored shafts, tank seams, etc.
	(ii)	Rubber repairs — rubber-based mixtures	Split, embrittled, rubber hoses. Damaged electrical insulation. Rubber flanges, etc.
24.	Adhesives	A wide range of natural and synthetic materials which are used to bond together other materials	Particularly useful for bonding dissimilar materials, dissimilar metals which constitute a corrosion couple, heat sensitive materials and fragile components
25.	Shrink insulation	An outer insulation sheath can be shrunk on to the existing insulation of a cable. The sheath is heat-shrinkable	In situ repair of damaged cable insulation where the cables are too difficult, or expensive, to replace

Index